SKILLING INDIAN AGRICULTURE

Author

Dr. R. M. Prasad is a retired faculty in the discipline of Agricultural Extension from Kerala Agricultural University (KAU). He served KAU for more than 32 years in various positions, including Associate Director of Extension.

Dr. Prasad worked as Training Specialist in two European funded projects in Kerala, viz., Kerala Horticulture Development Programme (KHDP) and Kerala Minor Irrigation Project (KMIP). He also served as a National Facilitator for Extension Reforms Programme spearheaded by MANAGE, Hyderabad. During his tenure as Senior Fellow in NIRDPR, Hyderabad, he was involved in the domain of Rural technology and Skill development. Dr. Prasad also served as Advisor, Department of Planning, Government of Meghalaya.

Dr. Prasad was a Member of the Research Advisory Committee (RAC) of two ICAR institutes, Indian Institute of Maize Research (IIMR), Ludhiana and Directorate of Cashew Research (DCR), Puttur, Karnataka. His fields of interest are Technology Transfer, S&T interventions, Convergence, Climate change and Skill development.

Dr. Prasad is the Life Member of Indian Society of Extension Education (ISEE), New Delhi and Indian Society for Training and Development (ISTD), New Delhi.

SKILLING INDIAN AGRICULTURE

Dr. R. M. Prasad

2026

Daya Publishing House®

A Division of

Astral International Pvt. Ltd.

New Delhi – 110 002

ISBN: 978-93-5919-484-4

Published by : **Daya Publishing House**®
A Division of
Astral International Pvt. Ltd.
– ISO 9001:2015 Certified Company –
4736/23, Ansari Road, Darya Ganj
New Delhi-110 002
Ph. 011-43549197, 8130496929
E-mail: info@astralint.com
Website: www.astralint.com

Printed at : **Replika Press Pvt. Ltd.**

FOREWORD

K. Jayakumar

This timely book by Prof. R. M. Prasad is a highly ambitious enterprise. Agriculture in India is a subject fraught with complexities and contradictions. The interests of the farmer and the consumer are at loggerheads. The interests of the middlemen and the farmer are at odds. Reforms in agriculture sector have always invited vehement responses and reactions. At the same time distress among farmers is on the rise while demand for food is growing. With climate change as an undeniable catastrophe waiting to reveal its true face, agriculture is beginning to feel unprecedented strains and conflicts. It is something that cannot be allowed to go unheeded. When agriculture wilts, a country will find it difficult to maintain order. In the twenty first century where technological solutions to many a concern of humanity are available, it should be the endeavour of all sensitive scientists and committed academics to explore new ways of empowering agriculture with new skills, attitudes and technology. This book on 'Skilling Agriculture' has to be contextualized in this macro situation and welcomed as the much-awaited impetus for modernisation and adaptation by skilling. "Increased pace of globalization, changed geopolitical and economic environment and technological changes, including digital advancements like artificial intelligence (AI) provide both challenges and opportunities for economic expansion, job creation and more possibly, restructuring. The transition to a more open and digital economy demands diverse sets of skills and availability of quality skills critical for a nation." The scope of the book is self-evident in these lines from the introductory chapter.

In fourteen chapters the author tries to capture the multidimensionality of the issue of skilling in agriculture. While skill and related concepts are explained initially, the book goes on to examine in fourteen chapters the issues and challenges in skill development, policy initiatives, skill development in other countries, agricultural extension services, livelihood and skill development for women, traditional skills and recognition of prior learning, agriculture skills under Artificial Intelligence, green skills and a host of most relevant topics. Every chapter purveys the issues comprehensively and attends to even minor details with scientific objectivity and precision. It is beyond the scope of this brief Foreword to (and beyond my competence) to examine every chapter critically. Suffice to say that every chapter has been designed with a clear comprehension of issues, eye for details and insistence on objectivity.

The relationship between education and skills in India is a problematic one. Though the National Education Policy 2020 and several new initiatives of the Union Government have rightly appreciated this gap and offer a number of skilling opportunities, the full impact of such initiatives is yet to be felt in the agricultural sector. There of course is a direct correlation between education and skills, agriculture and rural development and the SDGs. Agricultural production and productivity can stimulate employment and enhance incomes. The author underscores the virtuous cycle between food security, sustainable agriculture and education and skill training. However, apart from theoretical assumptions, the fact remains that much of the economy and much of the population are still rooted in traditional activities, attitudes and steeped in illiteracy. The structural barriers of Indian rural society need to be overcome if India has to achieve an agricultural miracle.

In such a context, skill development is not an easy task. Prof. Prasad makes an insightful appraisal of the Indian situation and challenges to skilling in these words: "It (skilling) is much more difficult to plan and deliver than any other level of education. Its demands tend to be unarticulated and changing. It faces a complex repertoire of competing interests and heterogeneous target groups, many of whom do not know exactly what they want to do after skill training." The basic constraint is one of perception; where farming is not viewed as a profession that can contribute to upward social mobility. In order to achieve social mobility, rural youth migrate out of rural areas and agriculture in search of better prospects in urban areas and often in tertiary sector, where they are forced to acquire 'enabling skills, key skills, core skills, essential skills, functional skills, life skills and enterprise skills.'

The prevalent skill gap in agriculture particularly is quite daunting with many young people attending schools without acquiring basic literacy skills. As a result they are unable to compete in the job market. There is widespread perception that vocational education in schools is an inferior option compared to general secondary or tertiary education. There is a demand-supply mismatch, the problem of obsolete curricula in vocational education, gender disparities, entrepreneur education gap and several such lacunae.

The book goes on to explain in great detail the positive and creative response of the Government in the recent past to address the issue of skilling. The NEP 2020 attempts to connect vocational and formal education by introducing flexible structures and emphasising experiential learning, enhancing technological innovation, encouraging international collaborations, and increasing public investment to improve educational infrastructure. There is the National Skill Development Mission (NSDM) Pradhan Mantri Kaushal Vikas Yojana (PMKVY) and the Centrally Sponsored State Managed (CSSM) component implemented by State Governments through State Skill Development Missions (SSDMs).

The book also discusses the area of emerging green jobs and the application of Artificial Intelligence (AI) in agriculture. This section provides substantial (and new) information on green jobs and allays some of the common misconceptions. The author clarifies that 'green jobs are decent jobs that contribute to preserve or restore the environment, be they in traditional sectors such as manufacturing and construction, or in new, emerging green sectors such as renewable energy and energy efficiency.' Green jobs are a priority for India, since the country is working towards its ambitious climate commitments to decarbonize the economy. Skill Council for Green Jobs (SCGJ) is one of the most recently launched initiatives of the Government of India aligned to the National Skill Development Mission. Green skills refer to the knowledge, abilities, and competencies required for individuals to contribute to a more sustainable and environmentally friendly society and economy.

This book is a tour de force unmatched in its scope and details. The complex issues of Indian agriculture and the challenges and urgency of skilling cannot be presented more adeptly than accomplished by this book. The book encompasses a wealth of information, rarely available in one place. While the canvas of the book is admittedly vast, the author has accomplished a fine balance between academic inputs and practical information. In marshalling diverse and abundant information, Prof. Prasad has shown remarkable restraint and wisdom and has kept the focus steady on the potential reader. This comprehensiveness and objectivity are the pillars of strength of this work. I hope the book succeeds in nudging the cardinal topic of skilling to nation's consciousness and public discourse. I compliment Prof. R M. Prasad for his painstaking work in producing a book which he can be proud of and the readers can feel abundantly rewarded.

K. Jayakumar

Sri K. Jayakumar is a bilingual poet, translator and author. Born on October 6, 1952 in Kerala, he has charted an distinguished career in the Indian Administrative Service (IAS) as well as literature. Author of forty-seven books in English and Malayalam, including twelve anthologies of poems and translations of Tagore, Kahlil Gibran, Rumi, and Kabir, Sri. Jayakumar has won several awards, including that of Kerala Sahitya Academy. Besides, he is a popular lyricist in Malayalam cinema, with over five hundred songs to his credit. He retired as the Chief Secretary of Kerala after an illustrious career in the State and the Central Governments, serving as Agriculture Production Commissioner, Home Secretary, etc in Kerala and as Joint Secretary in the Ministry of Culture. As the founding Vice-chancellor of Malayalam University, he played a pioneering role in higher education and excelled as an institution builder. He has also made a mark in scripting, filmmaking and painting. Currently Sri Jayakumar serves as the Director of the Institute of Management in Government, Kerala, which functions as the Administrative Training Institute for the State of Kerala.

PREFACE

The agricultural sector is considered the backbone of the Indian economy and plays a vital role in food security, job creation, and overall economic development, contributing significantly to GDP and international trade. However, this sector faces a huge challenge of ensuring a skilled workforce equipped to handle the complexities of modern agriculture.

The present-day agriculture cannot continue on traditional farming solely relying on inherited knowledge. Modern agriculture is a technologically advanced and data-driven industry with required skill sets. Also, the concerns around sustainability and environmental impact necessitate a workforce that understands and adopts best practices for conservation farming, including sustainable soil and water management, and responsible use of agrochemicals.

Though the State Agricultural Universities (SAUs), Krishi Vigyan Kendras (KVKs) and State Agricultural Departments in India play a crucial role in equipping the farmers with required knowledge and skill for efficient farming, various reports indicate that skill development in the farm sector is far from satisfactory. The performance of KVKs in providing skill trainings for farmers is also severely criticized. The over-utilization of KVKs for various activities by the state development departments as a means of offloading their own responsibilities is projected as a serious problem.

The establishment of Skill Universities in India is aimed at addressing the 'skill deficit' in various domains, including agriculture. It is expected that Skill Universities can play a vital role in developing agricultural skills by focusing on practical, industry-relevant training, preparing graduates for the modern workforce and addressing the needs of the agricultural sector.

Agriculture Skill Council of India (ASCI) is actively contributing to the skill development process in the domain of Agriculture, when country's agriculture is experiencing stagnant growth, exodus of quality manpower to other sectors, changing climate with increased variability in production parameters and transformations in international agriculture markets, too much subsidized challenging the competitiveness of Indian Agriculture. Though ASCI has taken upon itself the responsibility of transforming Indian Agriculture through developing the skills of country's manpower in emerging areas of agriculture, a lot needs to be done.

The present publication is designed against the also background to provide an overview of the state-of-the-art scenario of skill development in the farm sector in India. The publication has an important goal of highlighting the potential and scope of developing required skill sets in farmers for enhanced production and productivity and the need for blending the hereditary skills possessed by them, particularly in climate mitigation and adaptation.

I got an opportunity to offer an Elective course on 'Skill Development' for the students of Post Graduate Diploma in Rural Development Management (PGDRDM) in NIRDPR, while I was working as Senior Fellow in that institute. Sri. K N Kumar, IAS (Retd) who was serving as Deputy Director General (DDG) in NIRD guided and supported me in developing the course curriculum, which probably is the first course of its kind offered in the country. This book is dedicated to the first batch of PGDRDM students for our mutual learning process. I also got occasion to know more about skill development during field monitoring of the Placement linked skill development under SGSY in different states. In Meghalaya, I was also associated with the preparation of Meghalaya Skill Development Mission, thanks to Sri. K N Kumar Principal Secretary. The motivation for writing this book is the passion I continued trying to know and understand more about the developments in the Indian skill sector. I am thankful to Dr. Rasheed Sulaiman, who regularly shares the AESA blogs and related literature in Agricultural Extension and is a source of motivation for me. However, I personally feel that AESA should take more active interest in trying to enhance the skill sets of agricultural extension personnel for their effective delivery of services to the farmers.

After reading the book 'Credible: The Power of Expert Leaders' by Amanda Goodall (2023), I glanced through the Preface of the book and found an interesting observation, which I reproduce here, considered relevant for this book. 'People often point to robots, artificial intelligence, and other emerging technologies as the future of business. Yet, organisations also have much to gain by making one simple but profound adjustment: cultivating talented leaders with deep industry expertise and then training them with the requisite management and leadership skills. Such a transformation promises to make our workplaces happier and more productive.' I Sincerely wish that agriculture also must undergo the same transformation.

The book is organized under 14 chapters, including the introductory chapter, which introduces the significance and need for skill development for economic growth. Chapter 2 provides the conceptual clarity about skill and related terminologies, which helps in understanding the process of skill formation. The issues and challenges in the Indian skill sector are highlighted in the third chapter. The policy initiatives and associated developments in fostering the skill environment in India are discussed in the fourth chapter. The fifth chapter presents an overview of the skill structure in India and its governance mechanisms.

A brief attempt is made to review the skill development in other countries to provide a global perspective, which is presented in Chapter 6. Chapter 7 makes a detailed presentation of skill development in agriculture and allied sectors in the Indian context. In India, for all practical purpose, Agricultural Extension is considered as the vehicle for developing required skill sets among the farmers, which is discussed in Chapter 8. The farm women play an important role in the agricultural activities in India and considering their contribution, many skill development programmes have been formulated by the Government for enhancing their livelihood opportunities, which the Chapter 9 covers. The significance of Recognition of Prior Learning (RPL) in agriculture and its assessment is discussed in Chapter 10. There are many digital tools used in the present- day agriculture for enhancing the productivity and improving the efficiency of farmers, including use of generative AI in farming, about which Chapter 11 deals with. In the context of climate change and sustainability of natural resources, green skills emerge as a responsible option, which is covered in Chapter 12. Chapter 13 presents the scenario about future skills and discusses the importance of resilience in the volatile and uncertain environment. Chapter 14 offers concluding remarks and presents the way forward.

There is vast field of literature in the domain of skill development and is an emerging field. I had limited its scope to the field of agriculture. However, in this domain itself, it is quite possible that proper coverage is not done and there could be many lapses and omissions. Since this publication is a lonely endeavour, I remain solely responsible for the errors and lapses, if any. I earnestly request the readers to offer constructive suggestions and valuable feedback for improvement of this publication.

I acknowledge with gratitude the support and encouragement received from my family members, a real source of inspiration for active engagement.

I sincerely thank Sri. K. Jayakumar, IAS (Retd), former Chief Secretary, Government of Kerala and Director, Institute of Management in Government (IMG), Kerala for his impressive Foreword. He is a multi-talented personality and known public figure and I am very much obliged to him for his well-articulated, scholarly and thought provoking Foreword for the book, which definitely has added value to this publication.

I am thankful to Astral International Pvt Ltd, New Delhi for their co-operation, support and diligence in publishing this book within a short span.

R M Prasad

CONTENTS

ACRONYMS

AC-ABC	Agri Clinic and Agri Business Centre
ACF	Ambujam Cement Foundation
ADB	Asian Development Bank
AGI	Artificial General Intelligence
ANI	Artificial Narrow Intelligence
ASI	Artificial Super Intelligence
AI	Artificial Intelligence
AIC	Atal Incubation Centre
AICTE	All India Council of Technical Education
ALE	Adult Learning and Education
APEDA	Agricultural and Processed Food Products Export Development Authority
APCNF	Andhra Pradesh Community managed Natural Farming
ARYA	Attracting and Retaining Youth in Agriculture
AQTF	Australian Qualification Training Framework
ASCI	Agriculture Skill Council of India
ATARI	Agricultural Technology Application Research Institute
ATMA	Agricultural Technology Management Agency
AVC	Agricultural Value Chain
AWSAR	Augmenting Writing Skills for Articulating Research
B-ABLE	Basix Academy for Building Lifelong Employability
BANI	Brittle, Anxious, Non-linear and Incomprehensible
BIRD	Bankers' Institute of Rural Development
BMBF	German Federal Ministry of Education and Research
BMZ	German Federal Ministry for Economic Co-operation and Development

CAMM	Common Agenda on Migration and Mobility
CAPM	Certified Associate in Project Management
CATS	Credit Accumulation and Transfer System
CBC	Capacity Building Commission
CBSE	Central Board of Secondary Education
CoE	Centre of Excellence
CEDSI	Centre of Excellence for Dairy Skills in India
CEFMSI	Centre of Excellence for Farm Machinery Skills in India
CEHSI	Centre of Excellence for Horticulture Skills in India
CII	Confederation of Indian Industry
CSCM	Centrally Sponsored, Centrally Managed
CSSM	Centrally Sponsored, State Managed
CSCI	Construction Skill Council of India
CSR	Corporate Social Responsibility
DARE	Department of Agricultural Research and Education
DAESI	Diploma in Agricultural Extension Services for Input Dealers
DBT	Direct Benefit Transfer
DDU-GKY	Deen Dayal Upadhyay Grameen Kaushal Yojana
DFID	Department For International Development
DGT	Directorate General of Training
DPI	Digital Public Infrastructure
DST	Department of Science and Technology
DQR	German Qualification Framework
EAP	Enterprise Awareness Programme
EAS	Extension Advisory Services
EETP	Employability Enhancement Training Programme
EEI	Extension Education Institute
EGMM	Employment Generation and Marketing Mission
ENVIS	Environmental Information System
EPFO	Employees' Provident Fund Organisation
ERP	Enterprise Resource Programme
ESDP	Entrepreneurship and Skill Development Programme
ESIC	Employees' State Insurance Corporation

ESSCI	Electronic Sector Skill Council of India
FHEQ	Framework for Higher Education Qualification
FICCI	Federation of Indian Chamber of Commerce and Industry
FICSI	Food Industry Capacity and Skill Initiative
FMTTI	Farm Machinery Testing and Training Institute
FoCT	Friends of Coconut Tree
FPO	Farmer Producer Organisation
GAI	Generative Artificial Intelligence
GCC	Gulf Co-operation Council
GDP	Gross Domestic Product
GHG	Green House Gas
GSDP	Green Skill Development Programme
GSCI	Green Skill Council of India
HLPE	High Level Panel of Experts on Food Security and Nutrition
ICT	Information Communication Technology
ICAR	Indian Council of Agricultural Research
ICFA	Indian Chamber of Food and Agriculture
IMC	Industrial Motivation Campaigns
ICM	India Centre for Migration
IGVET	India German Vocational Education Training
IGnITE	Indo German Institute for Technical Education
IISC	India International Skill Centre
IIAASD	International Institute of Advanced Agricultural Skill Development
ILO	International Labour Organisation
IoT	Internet of Things
IPR	Industrial Policy Resolution
ISAP	Indian Society of Agribusiness Professionals
ISTD	Indian Society for Training and Development
ITI	Industrial Training Institute
IWM	International Workforce Mobility
JSS	Jan Shikshan Sansthan
KASE	Kerala Academy for Skills Education

KDISC	Kerala Development and Innovation Strategic Council
KVK	Krishi Vignan Kendra / Kaushalya Vardhan Kendra (Gujarat)
KSA	Kingdom of Saudi Arabia
KYP	Kaushal Yuva Programme
LEDP	Livelihood and Entrepreneurship Development Programme
LEMM	Labour and Employment Ministers Meeting
LIVE	Local Institute of Vocational Education
LLM	Large Language Modules
LXP	Learning Experience Platform
MANAGE	National Institute of Agricultural Extension Management
MASP	Maharashtra Agri Skilling Programme
MC	Micro Credentials
MDG	Millennium Development Goal
MDP	Management Development Programme
MEA	Ministry of External Affairs
MFPI	Ministry of Food Processing Industries
MKSP	Mahila Kisan Sashaktikaran Pariyojana
MeitY	Ministry of Electronics and Information Technology
MoLE	Ministry of Labour and Employment
MoRD	Ministry of Rural Development
MoU	Memorandum of Understanding
MSME	Micro, Small and Medium Enterprises
MUDRA	Micro Units Development and Refinancing Agency
NAARM	National Academy of Agricultural Research and Management
NABARD	National Bank for Agriculture and Rural Development
NAPS	National Apprenticeship Promotion Scheme
NASSCOM	National Association of Software and Service Companies
NATP	National Agriculture Technology Project
NCCF	Network for Certification and Conservation of Forests
NCL	National Commission on Labour
NCONF	National Centre for Organic and Natural Farming
NCVET	National Council for Vocational Education and Training

NDUW	National Database of Unorganised Workers
NEAT	National Educational Alliance for Technology
NEEM	National Employability Enhancement Mission
NGO	Non- Governmental Organisation
NDDB	National Diary Development Board
NFDB	National Fisheries Development Board
NHB	National Horticulture Board
NIAM	National Institute of Agricultural Marketing
NIELIT	National Institute of Electronics and Information Technology
NIESBUD	National Institute for Entrepreneurship and Small Business Development
NIFTEM	National Institute of Food Technology Entrepreneurship and Management
NILE	National Institute of Lifelong Education
NOS	National Occupational Standards
NPE	National Policy on Education
NPSD	National Policy on Skill Development
NPSDE	National Policy on Skill Development and Entrepreneurship
NRLM	National Rural Livelihood Mission
NSDC	National Skill Development Corporation
NSDF	National Skill Development Fund
NSFI	National Skill Foundation of India
NSSO	National Sample Survey Office
NVEQF	National Vocational Education Qualification Framework
NYK	Nehru Yuva Kendra
OECD	Organisation for Economic Co-operation and Development
OJT	On-the-Job Training
OSH	Occupational Safety and Health
PAGE	Partnership for Action on Green Economy
PDO	Pre Departure Orientation
PLFS	Periodic Labour Force Survey
PKVY	Pravasi Kaushal Vikas Yojana
PM-DAKSH	Pradhan Mantri Dakshta Aur Kaushalta Sampann Hitgrahi

PMKVY	Pradhan Mantri Kaushal Vikas Yojana
PMKK	Pradhan Mantri Kaushal Kendra
PMKUVA	Pramod Mahajn Kaushalya and Udyojakta Vikas Abhiyan
PM-VIKAS	Pradhan Mantri Virasat Ka Samvardhan
QP	Qualification Pack
RDSDE	Regional Directorate of Skill Development and Entrepreneurship
RKVY-RAFTAAR	Rashtriya Krishi Vikas Yojana- Remunerative Approaches for Agriculture and Allied Sector Rejuvenation
RMSA	Rashtriya Madhyamik Shiksha Abhiyan
RPL	Recognition of Prior Learning
RSETI	Rural Self Employment Training Institute
RYSE	Rapid Youth Success Entrepreneurship
RySS	Rythu Sadhikara Samsthan
SAA	State Apprenticeship Advisor
SAMETI	State Agricultural Management and Extension Training Institute
SAMVAY	Skill Assessment Matrix for Vocational Advancement of Youth
SCF	Skill Challenge Fund
SCGJ	Skill Council for Green Jobs
SDG	Sustainable Development Goal
SERP	Society for Elimination of Rural Poverty
SFAC	Small Farmer Agribusiness Consortium
SHREYAS	Scheme for Higher Education Youth in Apprenticeship and Skills
SIDH	Skill India Digital Hub
SIIC	Skill India International Centre
SIP	Skill India Portal
SMAE	Sub Mission on Agricultural Extension
SMAM	Sub Mission on Agricultural Mechanisation
SSC	Sector Skill Council
SSG	Skill Future Singapore
SSW	Specialised Skill Worker

STAR	Skill Training Assessment and Reward
SVC	Skill Value Chain
TASK	Telangana Academy of Skill and Knowledge
TP	Training Provider/ Training Partner
TPA	Third Party Aggregator
TVET	Technical and Vocational Education Training
UNDP	United Nations Development Programme
UNEP	United Nations Environment Programme
UNESCO	United Nations Educational, Scientific and Cultural Organisation
UNEVOC	International Centre for Technical and Vocational Education and Training
UNIDO	United Nations Industrial Development Organisation
VET	Vocational Education and Training
VUCA	Volatile, Uncertain, Complex and Ambiguous
WSC	World Skill Council

CHAPTER 1

INTRODUCTION

Skills and knowledge are the driving forces of economic growth and social development of any country. The economy becomes more productive, innovative and competitive through skilled manpower. The level of employment, its composition, and the growth in employment opportunities are critical indicators of the process of development in any economy. Increased pace of globalization, changed geopolitical and economic environment and technological changes, including digital advancements like artificial intelligence (AI) provide both challenges and opportunities for economic expansion, job creation and more possibly, restructuring. The transition to a more open and digital economy demands diverse sets of skills and availability of quality skills critical for a nation. We are now operating in a human powered economy, having transitioned from an industrial economy to a knowledge economy and to an economy powered by the hearts and minds of people, in short, humanity.

The employers in many developing countries report lack of skilled workers as a major and increasing bottleneck for their operations, affecting their capacity to innovate. The COVID-19 pandemic has brought the pre-crisis vision of equitable, relevant, and quality skills development into sharper relief, adding unforeseen urgency to the need for reform and highlighting the huge costs of inaction. In the context of climate change and digital farming, the need and importance of climate skills and digital skills for our farmers also become quite relevant and important.

In the present global context, where work is no longer defined by jobs, the workplace is no longer a specific place, many workers are no longer traditional employees, and human resources is no longer a siloed function, relevance of skill becomes still more important (Deloitee.com). Many of the technological changes like the emergence of generative artificial intelligence and the development of neurotechnology have already become an everyday reality, that create transformative shifts in how people think about work and workplace.

1.1. Skill development and economic growth

Skilled manpower fuels economic growth through efficient conversion of the resources. It can enhance productivity, and reduce poverty. The linking of skills and productivity not only benefit the enterprises and economy, but also facilitate different segments of the population, particularly the marginalized sections of the society to reap the benefits of economic growth through skill development. Skilling in relation to economic growth can be summed up as:

- Skills enable individuals to become more productive and generate higher incomes. Skilled manpower makes enterprises more productive and profitable. Skills enable national economies raise production and generate wealth. This is equally valid for both wage employment as well as self-employment.
- Developing appropriate skills for the informal sector need to be at the centre stage of pro-poor strategies. Though several reasons can be attributed for this, economic reason dominates, in terms of optimal utilization of the resources and quick returns to the economy due to the early transfer of skills. Skill development is usually linked to income benefits and hence is concerned with all the three kinds of poverty reduction strategies -alleviation, reduction and prevention. Improved skills and knowledge are essential for the poor to access decent work and add value to existing subsistence employment.
- Human capital in terms of skills and abilities also helps determine the amount of investment in physical capital in an economy. Skills and physical capital complement each other.
- The acceleration of technological change requires higher-skilled workers. New technologies are knowledge and skills-intensive and hence there is need to prepare the manpower to handle them efficiently.
- Demands for skills within enterprises depends on how the work is organized. Enterprises traditionally organize work along the assembly line, dividing each task into its most elementary components and thereby minimize the skills and training required. However, increased competition and the advances in information and communication technology (ICT) have prompted many firms to undertake fundamental changes in their internal organization and work practice to overcome the redundancy by reskilling and multi-skilling of the work force.
- Skills establish a pull factor for foreign direct investment. Alternatively, lack of human capital may deter foreign investment.
- Digitalization places a premium on skills. It is changing the products and services on the market and the processes through which they are provided, which has a qualitative employment impact in terms of the needed skills. The different digitalization technologies affect employment and skills differently.

1.2. Skills-Productivity-Growth Link

Skill development is an important driver to address poverty reduction by improving employability, productivity and helping sustainable enterprise development and inclusive growth. The increase in productivity can be attributed to varied reasons, skill development being the most important and necessary condition for productivity growth. The evidence from developed countries suggests that investments in education and skills helps economy to move to high growth sectors and break the low wages - low skill development syndrome.

Skills development is important due to its contribution in enhancing productivity at various levels.

a. Enhanced skills enable individuals to be more productive. Higher productivity leads to higher wages. When enterprises become more productive and profitable, national economies raise production and create wealth. This occurs due to complementarities between the physical capital and human capital on one hand and between the technology and human capital on the other.

b. The case of several countries in East Asia demonstrates the necessity as well the possibilities in skill development for developing countries. Apart from universalizing primary education, countries such as South Korea, Singapore and Malaysia have achieved formal skill training of the majority of their population as well high rates of employment of the trained workforce. The literature on the sources of sustained high growth in these economies have located this growth to the 'endogenous' processes that created a highly skilled workforce to adapt the technology to the requirements of the respective countries, do innovations at the level of the enterprises and meet ambitious targets in a short period of time. While many of these systems were put into place at an early stage of their contemporary development, they also undertook overhauling of systems in tune with the needs of a globalized economy. Thus, the need for the skill mix to change at different levels of development and in response to different degrees of openness was recognized and responded to quite early by these countries.

   The effort to view a change in the skill mix in fast changing economies has been recognized by South Korea, which has a Comprehensive Plan for Lifelong Education and Learning, which allows local communities and provinces to enter and exit the learning and training process in line with the requirements of the economy on the one hand, and with their own need for improvement. Similarly, Singapore has made major changes in its skill strategy through the introduction of a National Continuing Education and Training Framework and a Lifelong Learning Endowment Fund. In Malaysia, National Skills Development Act, 2006 was passed to promote the development and improvement of a person's ability needed for vocation through skill trainings.

c. In contemporary production systems, the demand for skills has changed to some extent because of extensive advances in the economies as a result of greater integration with the world economy. Increased competition and the introduction of ICT have prompted enterprises to make fundamental changes in their internal organization and work practices, necessitating the existence and acquisition of specific skills in the workforce.

d. Fast changing knowledge economies warrant new and higher competencies among the learners in the society. At the heart of these are the 'soft skills' such as communication, collaboration and network. These can be supplemented with other skills such as the ability to create, apply, share and distribute knowledge; to have the capacity to re-skill and re-train as the situation demands; and to become committed to continuous and lifelong learning.

1.1. Lifelong learning in Korea

The Lifelong Education Act of the Republic of Korea (2009) calls on the Ministry of Education to develop a comprehensive lifelong education promotion plan every five years at the national level. This national plan proposes mid and long-term policy objectives and a basic direction for lifelong education promotion. Since 2009, four plans have been enacted, resulting in improved access to lifelong education and the creation of the National Institute for Lifelong Education (NILE) as well as 17 local Lifelong Education Institutions. The digital age has led to considerable transformations in the cultural, social, economic and political environment. The fourth National Lifelong Learning Promotion Plan (2018–2022) sought to develop a flexible and high-quality lifelong learning system in response to technological innovation, the Fourth Industrial Revolution, and social changes like the population structure (aging society), an increase in life expectancy and changes in the professional world.

The goal of the Fifth Lifelong Education Promotion Plan (2023-2027) is 'promoting customized lifelong learning that everyone enjoys'. The plan aims to prepare individuals for changes brought about by the digital transformation and super aged society.

Source: https://www.uil.unesco.org>articles/republic-korea

1.2. Skills Future, Singapore

Singapore's educational policy has long-embraced the tenets of continuous learning – moving away from traditional 'front-loading' of education, to providing multiple entry points for the workforce to continually upskill and reskill during their working lives.

The nation-wide initiative is led by SkillsFuture, which brings together policy makers, academia, employers, unions and the workforce, for a cohesive and concerted effort towards preparing for the uncertainties of the future.

Skills Future Singapore (SSG) drives and co-ordinates the implementation of the national SkillsFuture movement, promotes a culture and holistic system of lifelong learning through the pursuit of skills mastery, and strengthens the ecosystem of quality education and training in SSG.

SSG's annual skills report provides details of priority economic sectors and skills needed to thrive in these sectors. The emerging domains in the Green Economy, the Digital Economy and the Care Economy, as well as Critical Core Skills that cut across multiple sectors are given higher importance in the report.

Source: https://www.skillsfuture.gov.sg/docs/default-source/initiatives/ssgskill

1.3. Skill and Sustainable Development

Poverty in developing countries is a global concern and poverty reduction is a recognized objective of international policy. The Millennium Development Goal (MDG) of halving poverty does not incorporate policy recommendations, either for the international community or for the national governments. In debates and discussions on poverty, the dominant position is that economic growth reduces poverty. Skill development is essential to enhance the capability of the rural poor, and by strengthening capability, progress can be made towards achieving the MDGs. The following issues highlight the potentially positive relationship between education and skills, agriculture and rural development and the MDGs.

- Much of poverty and hunger is concentrated in rural areas. There is also strong evidence for the direct impact of literacy rates on hunger and poverty. Agricultural production stimulates employment and improves incomes. At the same time, food security positively affects labour productivity, livelihoods and education. Thus, there is the possibility of a virtuous cycle between food security, sustainable agriculture and education and skill training.
- There is a clear effect of illiteracy on poverty, food security and nutrition and vice versa. Hunger, malnutrition and food insecurity erode cognitive abilities and reduce school attendance. Conversely, illiteracy and lack of education reduce earning capacity, and contribute directly to hunger and poverty.
- Women's role in agriculture, access to resources and skills is of great importance. The education of girls has positive effect on family's health, productivity and well-being. The gender gap is more pronounced in rural areas. Improved education, introduction of labour- saving technologies and better nutrition can significantly improve maternal health.
- It is clear that hunger and poverty compel over-exploitation of natural resources, thus compromising environmental sustainability. Providing skills for mitigating climate change can go a long way in ensuring environmental sustainability.

- A cross-sectoral focus such as that of education and skill for rural development necessarily promotes the multi-agency mobilization of resources for achievement of the MDGs.

Education, vocational training and lifelong learning are central pillars of employability, employment of workers and sustainable enterprise development within the Decent Work Agenda, and thus contribute to achieving the MDGs to reduce poverty. Skill development is the key in stimulating a sustainable development process and can contribute to facilitating the transition from the informal to the formal economy. Skill development is also essential to address the opportunities and challenges that emerge in transition economies in the context of globalization.

A low-skilled, low-productive, low-wage economy cannot rapidly reduce poverty. The vicious cycles of inadequate education, poor training, low productivity, poor quality jobs and low wages trap the working poor and exclude the workers without adequate and relevant skills from participating in economic growth and social development in the context of globalization. This also negatively affects the competitiveness of enterprises and their capacity to contribute to economic and social development.

Appropriate skill development will achieve the objective of decent work both by increasing the productivity and sustainability of the enterprise and thereby improving working conditions and the employability of the workers. Effective skill development requires a holistic approach, which encompasses the following features:

a. Continuous and seamless pathways for learning: start with pre-school and primary education that adequately prepares young people for secondary and higher education and vocational training; provide career guidance, labour market information and counselling as young women and men move into the labour market; provide workers and entrepreneurs opportunities for continuous learning to upgrade their competencies and acquire new skills throughout their lives.

b. Development of core skills: including literacy, numeracy, communication skills, teamwork and problem solving and other relevant skills as well as awareness of workers' rights and understanding of entrepreneurship as the building blocks for lifelong learning and capability to adapt to change

c. Development of higher order skills: professional, technical and human resource skills to capitalize on or create opportunities for high quality and high wage jobs

d. Portability of skills: based firstly on core skills to enable workers to apply knowledge and experience to new occupations or industries and secondly, on systems that codify, standardize, assess and certify skills so that levels of competence can be easily recognized by social partners in different labour sectors across national, regional or international labour markets.

e. Employability: for both wage work and self -employment, which results from a foundation of core skills, access to education, availability of training opportunities, motivation, ability and support to take advantage of opportunities for continuous learning and recognition of acquired skills. These are critical for enabling workers to attain decent work and manage change and for enterprises to adopt new technologies and enter new markets.

Breaking the vicious circle of poor education, low productivity and persistent poverty is crucial for promoting inclusive economic growth and decent jobs for all. Education, being an end in itself, is also a means to getting a decent job, especially for young people, while lifelong learning is indispensable in order to keep up with the changing skills needed for the labour market. It can also contribute to facilitating the transition from the informal to the formal economy. Skills development is also essential to address the opportunities and challenges to meet new demands of changing economies and new technologies in the context of globalization.

Skills development strategies are high on the priority list of almost all countries in all stages of development, for at least three reasons:

- **Skills matching :** to better forecast and match the provision of skills, both in terms of relevance and quality, with labour market needs;
- **Skills upgrading:** to adjust skills development programmes and institutions to technological developments and changes in labour markets so that workers and enterprises can move from shrinking, low-productivity economic sectors and professions to expanding, high-productivity sectors and occupations. Such adaptation requires permanent and regular re-skilling, skill upgrading and lifelong learning for workers to maintain their employability and enterprises to remain competitive; and
- **Skills for society :** to build up capabilities and knowledge systems within the economy and society which induce and maintain a sustainable process of economic and social development.

Skills development is of essential importance to the Decent Work Agenda and the promotion of productive employment. The organization of vocational and technical education and skills development has been at the core of the ILO's technical services. The ILO was instrumental in assisting the G20 in developing a common training strategy named 'A Skilled Workforce for Strong, Sustainable and Balanced Growth' adopted by the G20 leaders in 2011.

1.3 Decent Work

The International Labour Organization (ILO) defines decent work as "productive work for women and men in conditions of freedom, equity, security and human dignity". In general, work is considered as decent when:

- it pays a fair income
- it guarantees a secure form of employment and safe working conditions
- it ensures equal opportunities and treatment for all
- it includes social protection for the workers and their families
- it offers prospects for personal development and encourages social integration
- workers are free to express their concerns and to organise

The Decent work agenda was formulated by ILO in 1999. Later, at the 2005 UN World Summit, countries agreed to make employment and decent work for all member countries. Increasing employment and ensuring decent work for all are essential aspects of sustainable development. Quality employment and decent work conditions help reduce inequalities and poverty, and empower people, especially women, young people and the most vulnerable such as people with disabilities.

Source: https://international-partnerships.ec.europa.eu>policies

Recognising that well-integrated and adequately skilled workers benefit both the origin and destination countries, leaders of the G20 nations in 2023 have decided to develop skills-based migration pathways and achieve sustainably financed universal social protection coverage. To prepare for the future of work, the leaders of the world's top economies welcomed efforts to map global skill gaps and the development of a G20 policy to address skill gaps globally by strengthening national statistical data and extending the coverage of the ILO and Organisation for Economic Cooperation and Development (OECD) Skills for Jobs databases to G20 countries.

In 2015, the UN launched the 17 Sustainable Development Goals (SDGs). Adopted by 193 member states, the goals represent an important international step in setting humanity on a trajectory towards sustainable development. SDGs are complex, inter-related goals that require changing the way business is conducted. Skilling up is essential for development practitioners to achieve the ambitious SDGs. For example, climate mitigation and adaptation policies require changes in the way energy, transportation, water, urban development plans are formulated and implemented. The use of digital technology to enhance value creation in rural communities is similarly a skill set that needs to be encouraged. Achieving these development outcomes requires substantial reskilling and upskilling and capacity building support for all stakeholders, ranging from top policy makers to ordinary citizens, from elected representatives to service delivery providers, from civil society organizations to the general public and youth.

1.4. Approach to skill development

Education and skill development broadly comprises school education, higher education, and industrial/technical training including vocational training. In general, 'skill development' refers to the larger objective of 'equipping an individual with marketable skills'. However, in recent times skill development has been largely used in the context of technical/vocational training for the manufacturing/industrial or services sector.

There is an emerging consensus that skill development for the poor must be part and parcel of community based economic and political development. Communities need to mobilise around specific 'development alternatives' that address key political, social and economic constraints. Skill development should be driven by a 'people-centred pedagogy' which maximizes locally available skills and empowers the poor to learn for themselves. Support for skill development should be directly linked to the actual skill needs of the poor and, invariably, will need to be closely related to the on-going production activities.

The term "skills-first" is increasingly used by a wide range of organizations across the private and public sector to describe a new approach to talent management that emphasizes a person's skills and competencies rather than degrees, job histories or job titles with regard to attracting, hiring, developing and redeploying talent. By focusing directly on skills, rather than on how they have been acquired, a skills-first approach has the potential to democratize access to economic opportunities and pathways to good jobs for many more people than traditional approaches have done. Now, the focus is on skills-first economy, where the primary currency is not the degrees or certificates that individuals hold but the skills they possess and can apply in various contexts.

A number of recent developments have contributed to the emergence of skills-based talent management practices. These include labour shortages in various economies and sectors; the increase in remote work and changes in talent management practices during the COVID-19 pandemic; the growth of online learning and micro-credentials; and increasingly sophisticated use of artificial intelligence (AI) and big data for jobs and skills mapping. Having the right skills within an organisation also remains a critical concern for companies as they look to reinvent their business model, reduce costs and manage external economic pressures.

Despite the fact that India is a fast developing economy, various challenges related to skill development have forced the Government to reinforce that many more interventions need to be made to create more employment opportunities for the majority of Indians, to enable them to participate in the benefits of growth and to contribute to that growth. For this, the people must have education and skills that equip them for the labour market. One of the sources of the skilled workforce is the vocational education and training system. However, there is an awareness that the system has not been able to respond appropriately to the demands of the labour market. A key issue relates to the reforms/interventions needed to improve the effectiveness of the system. While on one side, India has to face the future with its changing realities, on the other, it must also deal with the nature of its established traditions and structures.

Changing realities: Globalization, competitiveness, climate change and knowledge economy are the changing realities. India's transition to a knowledge-based economy requires a new generation of educated and skilled people. Its competitive edge will be determined by its people's ability to create, share and use knowledge effectively. A knowledge economy requires India to develop knowledge workers who are flexible and analytical and who can be the driving force for innovation and growth. A skilled workforce with appropriate skill sets for climate change adaptation is also required.

Established realities: Demographic pressures and financial constraints are the well- established realities. Much of the economy and much of the population are still rooted in traditional activities and structures. Significant elements such as the cultural, social and political traditions of the country need to be retained and at the same time, new knowledge and skills have to be imparted.

1.5. Managing Skill Development

There is no systematic assessment of the demand for training in the unorganized sector of the economy. Most assessments of such demand have been made on behalf of segments of organized industry. These do not properly gel with the overall need to educate and train the predominantly informal sector labour force in India. However, almost all the available reports invariably highlight skill shortages and mismatches between demand and supply.

All countries that have succeeded in linking skills with productivity have targeted their skill development policy with three objectives:

(a) Meeting skills demand in terms of relevance and quality- Matching of skills supply and demand.

(b) Mitigating adjustment costs- Training in new skills and occupations help to insure against prolonged unemployment or underemployment and to maintain the employability of workers and the sustainability of enterprises.

(c) Sustaining a dynamic development process- This is focused on the strategic role of education and training policies in triggering and continuously fueling technological change, investment, diversification and competitiveness.

Skill development is arguably the most difficult process to manage in the whole spectrum of education and training. It is much more difficult to plan and deliver than any other level of education. Its demands tend to be unarticulated and changing. It faces a complex repertoire of competing interests and heterogenous target groups, many of whom do not know exactly what they want to do after skill training. Managing skill development becomes more difficult by the different forms of provision- formal plus informal training provision, modular, short term and long term. Skill training has a lower status than general and university education. In addition, it competes with the larger and more popular subsectors for financing. Above all, managing small scale, dispersed infrastructure is also difficult.

Developing standards is the first requirement for quality skill training. Customers and employers should play a major role in determining standards, which should be expressed in terms of competencies (output requirements), not inputs. Quality in skill development requires three elements- standards, sufficient inputs and measurement of outputs against the standards.

Rural people in general are the most disadvantaged in terms of access to services, including education and training, and are the worst served by infrastructure of various kinds. The situation becomes still worse in the case of backward and marginalized classes and regions. Skill development must be an integral part of broader employment and development strategy and should support the MDGs. Skill development is a priority for countries, particularly as they progress towards meeting the MDGs for basic education. Increasing number of entrants to the labour market (surging numbers of youth who have completed formal schooling, but lack practical skills useful in the labour market) fuel the concern for skill formation. Since much of poverty and hunger is concentrated in rural areas, skill development for the rural sector is very important from an economic perspective.

The surplus labour in the informal sector is the main challenge. Economies cannot generate sufficient wage employment opportunities for all the new entrants to the labour market. Thus, an increasing number of youth either are unemployed or work for themselves in the informal sector. Unfortunately, most skill training is geared toward wage employment and not the informal sector.

India's transition to a knowledge-based economy requires a new generation of educated and skilled people. Its competitive edge will be determined by its people's ability to create, share and use knowledge effectively. In a globalized economy, a large pool of skilled workers is indispensable. Skilled workers enhance the efficiency and flexibility of the labour market, reduce bottlenecks and are more easily absorbed into the economy, and their mobility is also high.

Skill development is a necessary priority for meeting at least two challenges- economic challenges and social challenges. Skills can contribute to economic growth and the international competitiveness of various sectors. It also leads to poverty reduction, employment generation, creation of social capital and heightened sense of self- esteem. Skill development also results in promoting key ILO values of decent work, social protection and equal opportunity of the workers which though intangible, is important.

Effective skill development systems help to connect education to technical training, technical training to labour market entry and labour market entry to workplace and lifelong learning. This way, skill development helps to sustain productivity growth and to translate the growth into more productive and better jobs.

A disconnect between the skills acquired and those needed in the labour market ultimately means that the skill training has not made any impact. Close linkage with the labour market is the first requirement of a successful system of skill development. Hence, the first criterion for effective skill development defines the gap between the skill demand and supply. In other words, skill development does not refer to the curricular or programmed source of education or training itself, but to the productive capacities acquired through the skill courses and programmes. The acquisition of these capacities is dependent on many factors, including good quality education and training and the presence of a supportive environment.

1.6. Market failures in Skill Development

With the view to fully exploit the growth potential of various sectors, there is a need to develop a skilled workforce that possesses the required competence and skills for taking up jobs generated in various sectors and thereby complement as well as participate in India's growth story. Skilling the workforce will enable youth to be prepared for the jobs generated in the economy due to the high economic growth and will also act as an enabler of that growth. However, there are various market failures observed in skill development.

Where the free market fails to deliver efficient economic outcomes, this is termed as 'market failure'. Market failures are of four different kinds: Externalities, Asymmetric information, Market power and Public goods (Kelkar and Shah, 2019).

Externalities could be either positive or negative. Increased economic products and innovation and entrepreneurship are considered as positive externalities, while skill mismatch, brain drain and wage depression are negative externalities. It is well documented in the literature that skill development is an area where markets do not typically deliver optimum volumes of skilled manpower that the economies need. This is because of 'externalities' in skill and training provision. For instance, if a private company undertakes skilling of its workers, it might not be able to reap the benefits fully if the worker leaves the company. Another company, on the other hand, might benefit without bearing the cost of training if the skilled worker joins the company. This results in the 'under provision' of skills if it is entirely left to market forces. It is argued that the under-provision of formal skill training observed in small and micro enterprises is either due to 'ignorance' on the part of employers or 'market failure' due to a variety of supply and demand side features such as their short- term perspective which prevents the recognition of the long-term benefits of training.

Skills are merit goods that necessitate public funding and government intervention to create a skilled work force. The dual purpose of skill development is personal growth and securing employment. It is to be clearly examined and understood as to what and how much the government can do in the sphere of skill development. It would be neither feasible nor desirable for the government to attempt the skill training of entire youth in diverse trades across various regions of the country with the general aim of helping them secure a livelihood. The role of private sector and civil society in this regard is equally important.

While the domain of skill development is considered as one where public or collective institutions become necessary due to externalities mentioned above, where the basic stimuli will come from external sources in signaling the extent and type of skill requirements, whether skill development should be demand led or supply led become extremely important, particularly when the financing aspect is considered. The existing systems of formal skill provisioning in most developing countries are found to cover only a very small proportion of enterprises, creating large supply-demand mismatches, resulting in skill shortages. The issue of matching skills to markets has led to a call for dismantling existing system of skill provisioning in many countries and an overhaul of technical and vocational education training (TVET) systems.

1.7. Skills in Agriculture: Unpacking social meanings

Agriculture remains an important part of the livelihood portfolios of rural youth, a fall-back that helps them cope both with seasonality and the insecurities within alternate activities (HLPE, 2021). However, farming is not viewed as a profession that can contribute to upward social mobility.

Against this perception, the youth evolve diverse strategies to learn skills, refashion their identities and attain social mobility. This is achieved in some cases through migration, as the harsh conditions of work required for successful migration command respect within communities (Rao and Hossain, 2012).

Many young men, however, now use migration as a stepping-stone to earn capital, which can be invested in 'modernizing' their family occupation through acquisition of new skills and technologies. In the context of the fishing industry in Tamil Nadu, Rao and Manimohan (2020) found that young men, while migrating for short periods of time to raise capital, prefer to settle back in their communities as 'boat-owners' rather than 'fishing labour'. They invest in modern technologies like GPS and fish finder devices to make their occupation more technologically savvy and construct their identities as 'modern'. In terms of learning and upgrading skills in agriculture, especially post the Green Revolution in the 1960s, Agricultural Extension has played an important role.

One of the most significant features of the Indian labour market is a slow and steady transition of the workforce away from agriculture and into the non-farm sectors. The share of manufacturing employment was stagnant, at around 12-14 per cent. The increase in non-farm employment was absorbed by construction and services. After 2019, this slow transition reversed due to the pandemic, with a rise in the share of agricultural employment as well as an increase in the absolute size of the agricultural workforce.

During 2000-19, there was a shift in employment from low-productivity agriculture to relatively higher-productivity non-agriculture sectors. However, this transition slowed and then reversed between 2019 and 2022. This surge can be attributed to individuals returning to subsistence activities in agriculture due to the lack of work opportunities outside the agriculture sector exacerbated by the pandemic-related economic slowdown.

Employment in India is predominantly self-employment and casual employment. Nearly 82 per cent of the workforce engages in the informal sector, and nearly 90 per cent is informally employed. Due to the nature of employment growth since 2019, the share of total employment, which is in the informal sector and/or in informal employment, increased.

Due to increasing mechanization and capital use, the employment generation in India has become more and more capital-intensive, with fewer workers employed between 2000 and 2019 than in the 1990s. The skill intensity of employment in industry and services increased during this period, which was contrary to the labour market needs of the country.

Employing youth is a major challenge because of their emerging numbers combined with the lack of gainful employment opportunities in other non-farm sectors. Hence, ways to create various job roles within the agriculture and food sector can create a win-win situation for both the people seeking employment as well as the sectors in terms of higher productivity and growth. This will also contain jobless migration from rural to urban areas, a trend that has serious implications on the socio-economic well-being of the people.

1.8. Skills for Agricultural Transformation

Agricultural transformation is much more than changes in farming practices. It is about catalyzing transformation of a country's rural economy. Transforming a country's agriculture sector can create jobs, raise incomes, reduce malnutrition, and kick-start the economy on a path to middle-income growth.

Indian agricultural sector broadly comprises of farming (crops and horticulture) and forestry, livestock (milk, eggs, meat), and fisheries. Additionally, the sector also impacts consumption and production dynamics in non-agricultural segments, such as consumer products, retail, chemicals, and e-commerce. The economic interlinkages make the agricultural sector central to India's economic output and growth potential.

Agriculture is a major contributor to the Indian economy, accounting for around 18 percent of the country's GDP and providing employment for around 50 per cent of the population. But farming in India faces many challenges, some of which are:

- The average farm size in India is small, with around 85 percent of farms being less than two hectares in size, which affect profitability of farming.
- Unsustainable farming practices, such as overuse of pesticides and fertilizers, have resulted in soil degradation and reduced crop yields in many parts of India.
- Small farmers in India often have difficulty accessing markets to sell their products, mostly depending on middlemen, which limit their income and economic growth.
- India is particularly vulnerable to the impacts of climate change, such as changing weather patterns and increased frequency of extreme weather events, which can reduce crop yields and make farming more difficult.
- Many farmers in India are facing labour shortages as a result of immigration policies and a lack of interest of youth in farming as a career.
- The cost of production including the cost of inputs such as seeds, fertilizer, and farm implements is increasing which render it difficult to have farming as a profitable enterprise.

To address these challenges, there is transformation needed in agriculture related to:

- Raising agricultural productivity per unit of land: This needs to be the main engine of agricultural growth as virtually all cultivable land is farmed. Water resources for irrigation are also limited. There is need for increasing yields, productivity, diversification to higher value crops, and developing value chains to reduce marketing costs.
- Poverty alleviation through a socially inclusive strategy that comprises both agriculture as well as non-farm employment. Nutritional security is also to be considered.
- Application of emerging technologies to ensure that the agricultural growth responds to food security needs and addresses the climate and market related risks in farming.

In the case of agriculture, first and foremost, a need is to be felt for "transformation readiness." A country's institutional framework, governing mechanisms, and political environment significantly influence the readiness for accelerating an agricultural transformation. There are also drivers related to delivery mechanisms, which focuses on what is needed to translate the national agricultural plan into on-the-ground impact. The extension and advisory services of the country are also moving in this direction. Capacity building of the various actors involved in development of the sector also plays a crucial role.

Successful agricultural transformations have focused on the farming household, providing opportunities for farmers to earn a better income. For some, transformation means raising farm productivity or shifting the mix of production to include higher-value crops and livestock. For others, the right choice will be to do less farming and take advantage of employment options off the farm. Some others view agriculture as a business enterprise. Whatever may be the goal, agriculture is in a state of transition. The farmers have to be empowered to do their jobs better through appropriate capacity building programmes, including providing necessary skill sets.

As compared to other developed and developing countries, India has a unique window of opportunity for another 20-25 years called the "demographic advantage/dividend". If India is able to skill its people with the requisite life skills, job skills or entrepreneurial skills, in the years to come, the demographic advantage can be converted into the dividend wherein those entering labour market or are already in the labour market contribute productively to economic growth both within and outside the country.

References

Dayananda, K. C. 2023 Skill development and Economic growth, IOSR Journal of Humanities and Social Sciences, Vol 28, Issue 8, Series 2 (Aug 2023) 36-41

Dubey, Madhuri 2021 Sustainability in Skill development (Skill talk, Jan 27, 2021) https://nationalskillsnetwork,in>sustainability-in-skill

Ganguly, K., Gulati, A, von Braun, J (2019) Skill development in Indian Agriculture and Food processing sectors: A scoping exercise, ZEF Working Paper Series, No.183, https://www.econstor.eu>bitstream

HLPE Report 2021 Promoting youth engagement and employment in Agriculture and Food systems, Committee on World Food Security, https://openknowledge.fao.org>core>content

ILO 2011 A Skilled Workforce for Strong, Sustainable and Balanced growth, Feb 2011, https://www.ilo.org>publications>skill-workforce-strong

ILO 2023 Towards lifelong skills development for decent work, https://www.ilo.org>resources>article>towards-lifelong

ISB 2024 A policy roadmap for Skilling Indian youth, https://execed.isb.edu>policy-perspectives>articles>a-policy

Kelkar, Vijay and Shah, Ajay 2019 In Service of the Republic: The Art and Science of Economic Policy, Penguin Random House India, Gurgaon

Rao, N and M I Hossain 2012 "I want to be Respected: Migration, Mobility and the Construction of Alternate Educational Discourses in rural Bangladesh, Anthropology and Education Quarterly, 43 (4): 415-428

Rao, N and R. Manimohan 2020 (Re) Negotiating gender and class: new forms of co-operation among small scale fishers in Tamil Nadu, In UNRISD Occasional Paper 11, 32 https://www.unrisd.org/UNRISD/website/document nsf. publicity

CHAPTER 2

SKILL AND RELATED CONCEPTS

2.1. Definition and Meaning

There are three domains of learning- cognitive, affective and psychomotor. Cognitive domain refers to knowledge, affective is related to attitude and psychomotor refers to skills. Usually, skill is used to denote a level of performance, in the sense of accuracy and speed in performing particular tasks. Skill was characterized by Pear (1927) being concerned with the quantity and quality of motor outputs. It is considered as the integration of well-adjusted muscular performances.

Welford (1968) defined skill as a combination of factors resulting in 'competent, expert, rapid and accurate performance'. Though Welford's work initially focused on perceptual-motor performance, his later research on skilled performance (Welford, 1976) has increasingly taken into account broader cognitive skills such as problem solving and decision making. The acquisition of skill and the demonstration of 'skilled performance' involve a combination of underlying perceptual, cognitive and motor skills.

Proctor and Dutta (1995) define skill as goal-directed, well-organized behaviour acquired through practice and performed with economy of effort. In this definition, each element is important, as below:

- Skill develops overtime, with practice eg: driving skill.
- It is goal directed in response to some demand in the external environment eg: passing a driving test for driving license.
- It is acquired when components of behaviour are structured into coherent patterns eg: knowing how to use accelerator and clutch while driving.
- Cognitive demands are reduced as skill develops eg: driving without much concentrated and deliberate efforts with more practice.

Proctor and Dutta (1995) distinguish perceptual skills, response selection skills, motor skills and problem-solving skills. Perceptual skills are concerned with the ability to make distinctions and judgements; more complex situations require attentional control for processing but many tasks that initially require attention become automatised. Skill in selecting the appropriate response can be developed with practice.

DFID (2007) defines skill development as the acquisition of practical competencies, know-how and attitudes necessary to perform a trade or occupation in the labour market.

The Planning Commission's Committee on India Vision 2020 indicated that skill can be perceived as the ability to direct human energy efficiently to achieve desired goals. It is one of the attributes that generates knowledge resources, the others being technology, organization, information and education.

NITI Aayog (2015) considers skill development as an important driver to address poverty reduction by improving employability, productivity and helping sustainable enterprise development and inclusive growth.

Skill has been defined as an ability and capacity developed through thoughtful, cautious, systematic and persistent effort to efficiently and successfully accomplish multi-faceted activities or job functions involving ideas (intellectual skills), resources like computer software or hardware (specialized and practical skills), and/or people (social interactive skills) (www.businessdictionary.com/definition/skill.html).

An attempt is made to list out the definitions of skill collected from various sources, as below:

- An ability and capacity acquired through deliberate, systematic and sustained effort to smoothly and adaptively carry out complex activities or job functions involving ideas (cognitive skills), things (technical skills) and/ or people (interpersonal skills).
- An individual's ability to perform a specific task or solve problem at a high level of proficiency.
- A skill is an ability and knowledge that enables a person to do something well.
- One's ability to use acquired knowledge to perform a specific action and produce concrete results.
- An ability to carry out a task with pre-determined results, often within a given amount of time, energy or both.
- The ability to use one's knowledge effectively and easily in execution or performance of any physical tasks.
- The ability and promptness to transfer and apply knowledge and other resources to known and controlled situations.

- Skill is a learned ability to do something in a competent way.
- Specific ability that shows the application of knowledge acquired previously in specific settings and circumstances.

Skills Development can be operationally defined as the process of acquiring, improving, and enhancing a wide range of skills and competencies. It involves the deliberate and continuous effort to build knowledge, expertise, and capabilities in specific areas to meet the demands of one's work or to pursue personal and professional growth.

The current definition of "skill" generally refers to the ability to perform a task with expertise and proficiency, often gained through practice, learning, or experience. It involves both knowledge and the practical application of that knowledge.

Some common synonyms of skill are **artifice, art and craft**. While all these words mean "the faculty of executing well what one has devised," skill stresses technical knowledge and proficiency. While in some cases nearly identical to skill, art implies a personal, un-analyzable creative power. While the synonyms artifice and skill are much close in meaning, artifice suggests technical skill, particularly in imitating things in nature. The words craft and skill can be used in similar contexts, but craft may imply expertness in workmanship.

In practical sense, the term used is Marketable skill, commonly understood to refer to any skill/expertise/ability that has market, has the potential of being utilised for generating income/employment. According to NSSO, any marketable skill, whether acquired through formal or informal means, irrespective of whether it is marketed or not, whether the intention is to market it or not, is considered skill.

Skill development is the bridge between education, training and production/ work environment.

Education is about developing a person's potential, helping to discover the abilities that he/she has and giving him/her the tools to use them. Education imparts knowledge, skills and attitude for application to life. It is a formal phase of learning, which is a broad concept.

Training refers to imparting specific practical skills, to be used for specific and limited purpose. Training imparts skills, knowledge and attitude for direct application to a particular task or job. It is a specific concept.

Skill development relates work environment. Development is a larger canvas which envisages modification of behaviour and personality. It is a multi-dimensional concept, as most jobs require a combination of skills for adequate performance, ranging from physical abilities to cognitive and interpersonal skills.

2.2. Skills viewed by different disciplines

Various disciplines view skill in their own perspective. Of the different disciplines, Economics, Sociology, Education and Psychology are considered to be more influential. Economists define skill as an individual's human capital. They view training as major process of human capital formation primarily under two micro-level perspectives. First, they consider training as a personal investment into human capital. In the second perspective, they look at training from the point of view of the employer and of the economy as a whole.

In Sociology, skills are recognized as institutionally defined and often credentialed bundles of work-related qualifications, frequently cast in the form of occupations and professions. Sociologists are concerned with institutional frameworks of skill acquisition, as well as their variation across time and among societies.

Psychology considers skills as a component of behaviour and skill formation as part of individual learning process. Accordingly, this discipline brings to the table a long-standing expertise in general and universal theories about learning.

Educationists view skill as an acquired trait after getting exposed to varied learning experiences and learning situations.

Though various disciplines, including psychology, education, and economics, have defined and analyzed skill, there does not exist a single, universally recognized authority behind the definition. Influential researchers in cognitive psychology, such as **Daniel Kahneman** and **Anders Ericsson,** have contributed significantly to the understanding of skill acquisition and development. Since the disciplines in the field of skill development are segmented, a cross disciplinary framework is necessary for understanding the skill formation and development process. Such a framework is currently lacking in the strict sense.

2.1 Anders Ericsson's contribution to skill development

- Ericsson's research focuses on how individuals develop exceptional skills and expertise through deliberate practice, characterized by focused attention, feedback and repetition.
- Ericsson questioned the notion of innate talent as a primary driver of expertise and emphasized the crucial role of practice over years in attaining exceptional performance.
- Ericsson's work emphasizes the significance of structural and focused training approaches to foster skill development and improve performance.
- Ericsson's work has contributed to the popularization of deliberate practice as a key strategy for achieving skill in various domains.

Source: https://www.journalofexpertise.org>article

2.3. Types of Skills

Proctor and Dutta (1995) distinguished perceptual skills, response selection skills, motor skills and problem- solving skills.

- Perceptual skills are concerned with the ability to make distinctions and judgements.
- Skill in selecting the most appropriate response can be developed with practice, reaction time is affected by a number of alternatives.
- Motor skills are the manual aspects of performance such as speed and accuracy of physical movements or dexterity.
- Problem solving skills dependent on intellect and mental processes are acquired and developed through practice.

Skills are also classified as:

- Practical skills, Information skills, People skills and Performance skills
- Spatial skills, Logical-mathematical skills and Kinesthetic skills. Spatial skills apply to the perception, classification and identification of objects; Logical-mathematical skills address the individual and collective properties of objects and relate to numeracy; Kinesthetic skills concern the perception and use of one's body.
- Basic skills, People skills People related skills, Thinking skills, Personal skills, Business skills and Community skills.
- Hard skills (technical/administrative procedures related to core business of an organization) and Soft skills (needed for everyday life; how people relate to each other (communication, collaboration, team work).

Though there are many types of skills, the types often seem to overlap.

2.4. Employability Skills

Employability refers to a person's capacity of gaining initial employment, maintaining employment and obtaining new employment, if required (Hillage and Pollard, 1998). Employability is not just about vocational and academic skills. For individuals, employability depends on:

- Their assets in terms of knowledge, skill and attitude they possess.
- The way they use and deploy those assets.
- The way they present them to employees.
- The context (personal, labour market) within which they seek work.

Employability skills are those basic skills necessary for getting, maintaining and doing well on a job. Unlike occupational or technical skills, employability skills are generic in nature, rather than job specific. The different terms for skills similar to employability skills used in various contexts are: **generic skills, enabling skills, key skills, core skills, essential skills, functional skills, life skills and enterprise skills.**

The use and popularity of concepts of generic skills has increased globally. No matter where they have been adapted, or how they have been named, the conceptualization of skills shares a common purpose. It seeks to establish the basis for recognizing an important set of skills that support the successful accomplishment of the task-based activities central to any job role.

While generic skills may have contextualized applications unique to a work place and job role, it is important also to understand that they are highly transferable. A generic skill learned or applied in one workplace will also be applicable in another. The success of an individual in a new job role is, in part, based on his ability to draw on previous experiences and relate them to the present situation.

Table 1. Elements commonly found in generic skill listings

Type of skill	Attributes
Basic /Fundamental skills	Literacy, Numeracy, Technical skills
People-related skills	Communication, Interpersonal relations, Teamwork, Customer service skills
Conceptual/Thinking skills	Information processing, Problem solving, Analytical skills, Creative thinking, Planning skills
Personal skills	Being responsible, resourceful and flexible; Self-management skills
Business skills	Innovation skills, Enterprise skills
Community skills	Civic and citizenship knowledge and skills

Ref: Gibb, Jennifer (2004) Generic skills in vocational education and training, National Centre for Vocational Education Research Ltd, Australia.

The report 'Employability Skills for the Future' published by the Business Council of Australia and the Australian Chamber of Commerce and Industry (2002), outlined an 'Employability Skills Framework, which identifies eight employability skills. They are:

- Communication skills, which contribute to productive and harmonious relations between employees and customers.
- Teamwork skills, which contribute to productive working relationships and outcomes.
- Problem solving skills, which contribute to productive outcomes.
- Initiative and Enterprise skills, which contribute to innovative outcomes.
- Planning and Organizing skills, which contribute to long term and short-term strategic planning.
- Self- management skills, which contribute to employee satisfaction and growth.
- Learning skills, which contribute to ongoing improvement and expansion in employee and company operations and outcomes.
- Technology skills, which contribute to effective execution of tasks.

2.2 Employability Skills for CTS

The employability skills for Craftsmen Training Scheme (CTS) in India under competency based curriculum was revamped in the 16th National Skills Qualifications Committee (NSQC) of NCVET held in February, 2022. It was decided to revamp the existing employability skills to cater to the dynamic industry needs by imparting the trainees with adequate employability skills like digital skills, online payment, communication, entrepreneurship, marketing etc. Accordingly, Employability Skills syllabus of 60 hrs. duration for all six months trades and 120 Hrs. duration for all one year and first year of two-year CTS trades is designed in collaboration with National Skill Development Corporation (NSDC).

Source: www.cstaricalcutta.gov.in

Global mega trends such as the rising role of technology, climate change, demographic shifts, urbanization, and the globalization of value chains are changing the nature of work and skills demands. To succeed in the present labour market, one needs a comprehensive skill set comprising of:

1. **Cognitive skills,** which encompass the ability to understand complex ideas, adapt effectively to the environment, learn from experience, and reason. Foundational literacy and numeracy as well as creativity, critical thinking, and problem-solving are cognitive skills.
2. **Socio-emotional skills,** which describe the ability to navigate interpersonal and social situations effectively, and include leadership, teamwork, self-control, and grit.
3. **Technical skills,** which refer to the acquired knowledge, expertise, and interactions needed to perform a specific task, including the mastery of required materials, tools, or technologies.
4. **Digital skills,** which are cross-cutting and draw from all of the above skills, and describe the ability to access, manage, understand, integrate, communicate, evaluate, and create information safely and appropriately.

The development of skills can contribute to structural transformation and economic growth by enhancing employability and labour productivity and helping countries to become more competitive. Investment in a high-quality workforce can create a virtuous cycle, where relevant and quality skills enable productivity growth and foreign direct investment, which result in more and better jobs for the current workforce and more public and private investment in the education and training system. This, in turn, increases the employability and productivity for both the current and future workforce.

2.5. Life Skills

Life skills are abilities for adaptive and positive behaviour that enable humans to deal effectively with the demands and challenges of life. This concept is also termed as psycho-social competency. Life skills improves the quality of life of individuals and are also important to employers, who need workers mentally stable and well equipped to handle challenges and responsibilities not included on the job description.

Though there are many life skills, the ten core life skills laid down by WHO are:

1. Self-awareness
2. Empathy
3. Critical thinking
4. Creative thinking
5. Decision making
6. Problem Solving
7. Effective communication
8. Interpersonal relationship
9. Coping with stress
10. Coping with emotions

Self-awareness includes recognition of the self, our character, our strengths and weaknesses, desires, and dislikes. Developing self-awareness can help to recognize when we are stressed or feel under pressure. It is often a prerequisite to effective communication and interpersonal relations and for developing empathy for others.

Empathy is the ability to imagine what life is like for another person. Without empathy, our communication with others will result only as one-way traffic. Empathy can help us to accept others, who may be very different from ourselves.

Critical Thinking is the ability to analyse information and experiences in an objective manner. It can contribute to work by helping us to recognize and assess the factors that influence attitudes and behaviour, such as values, peer pressure, and the media.

Creative Thinking is a novel way of seeing or doing things characteristic of four components, viz., fluency (generating new ideas), flexibility (shifting perspective easily), originality (conceiving of something new), and elaboration (building on other ideas).

Decision-making helps us to deal constructively with decisions about our lives. It can teach people how to effectively make decisions about the actions in relation to the assessment of different options and, what effects these different decisions are likely to have.

Problem-solving is the process of identifying and resolving conflicts or problems. It involves breaking a problem into its component parts, thinking about possible solutions, and then choosing the best one. It is an important skill for everyone since it helps to identify and solve problems in our lives, work, and communities.

Interpersonal relationship skills help us to relate in positive ways with the people we interact with. This may mean being able to make and keep friendly relationships, important to our mental and social well-being. It also means keeping good relations with family members, considered an important source of social support.

Effective communication means that we are able to express ourselves, both verbally and non-verbally, in ways appropriate to our cultures and situations. This means being able to express opinions and desires, and also needs and fears.

Coping with stress means recognizing the sources of stress in our lives, recognizing how it affects us, and acting in ways that help us control our stress levels by changing our environment or lifestyle and learning how to relax.

Coping with emotions means recognizing emotions within us and others, being aware of how emotions influence behaviour, and responding to emotions appropriately. Intense emotions like anger or sadness can have negative effects on our health if we do not respond appropriately.

2.6. Multidisciplinary skills

Multidisciplinary skills relate to abilities that can facilitate the process to find solutions to complex problems from several perspectives. Multidisciplinary skills involve drawing on knowledge and expertise from various fields to tackle complex problems, fostering collaboration, critical thinking, and creative problem-solving.

In the rapidly evolving new economy, disruptions caused by technology have reshaped the job market, requiring professionals to continuously upskill and adapt. To succeed in this dynamic landscape, individuals must embrace the habit of life-long learning and cultivate a multidisciplinary profile with a diverse skill set.

The new economy favours professionals with a multidisciplinary profile and a diverse skill set. The ability to leverage knowledge and skills across different domains is highly valuable in today's rapidly changing job market.

2.7. Upskilling, Reskilling and Cross-skilling

Upskilling entails skill development to enhance the expertise of employees in the existing job role without changing their jobs. It involves training existing staff and developing their skills further, so that they are better equipped to meet new challenges. It typically involves training around new technology, trends, or standards and can help a workforce become more effective by enhancing critical competencies. For example, organisations are upskilling their workforce to use advanced tools and concepts required in their current roles to develop in-house experts in a particular technology/domain.

Reskilling refers to building employees' capabilities to prepare them for different job roles as their current skill sets become obsolete due to technological advancements. It involves the process of teaching employees entirely new skills, aligning them with new positions or mandates. It may involve significant shifts in the abilities of employees but can help an organization bridge its skills gaps, the current workforce is lacking. For example, organisations have reskilled their analysts and data engineers to be proficient in skills required to perform AI-related work.

Cross-skilling involves training existing staff in skills outside their current areas of expertise, so they can take on multiple roles that complement each other. This can result in a more flexible workforce and an improved ability to handle unexpected situations. For example, a graphic designer may learn about website development to understand how their work fits the website's overall theme.

2.8. Microlearning: A Key Enabler of Upskilling and Reskilling

Microlearning is an educational strategy that focuses on learning new information in small units. This approach to learning breaks down topics into short-form, standalone units of study that the learner can view as many times as necessary, whenever and wherever they need.

Microlearning gives learners access to short bursts of new information that is engaging and also digestible, as opposed to one large chunk of information that might be difficult to comprehend and remember. The convenience of microlearning, from both the learner's and the educator's point of view, has made this type of instructional delivery popular in corporate learning environments. This approach can improve knowledge retention while also empowering employees by letting them build new skills without having to take time away from their jobs to attend training.

Microlearning instructional modules are designed to be consumed in about five minutes and address one specific skill or knowledge gap topic. The learners can complete their training when it best suits their schedule.

Microlearning has emerged as a powerful tool in the pursuit of lifelong learning. The short, focused, and easily accessible learning modules delivered can cater to specific skill development needs and offer several advantages over traditional methods of learning, such as long-term courses or degree programmes. The salient features are:

- **Flexibility and convenience**

Professionals can access learning materials anytime and anywhere, fitting their studies into their busy schedules. This enables them to engage in continuous learning without disrupting their work or personal commitments.

- **Targeted learning experiences**

Professionals can focus on specific skills or competencies that align with their career goals or the evolving demands of the job market. This personalized approach optimizes the learning process, ensuring that efforts directly contribute to professional growth and employability.

- **Just-in-time learning**

Professionals can acquire new knowledge and skills on-demand, exactly when they are needed. For instance, they can quickly learn about a new programming language, digital marketing technique, or data analysis tool before applying it to their work. This agile learning approach ensures that professionals are up-to-date with the latest industry trends and technologies.

2.9. Skill training, Skill development and Skill Management

Skill training refers to the process of acquiring or enhancing specific abilities, knowledge, or competencies to perform tasks effectively. It includes planned educational activities meant to foster transferable abilities related to a specific career path, job function, or area of interest. The purpose of skill training is to nurture talent and thus help people to upskill with more than one specialisation. Skill training remains a valuable investment for individuals and organizations seeking to enhance performance, competitiveness, and overall success.

Elements of skill training

Several elements are involved in successfully imparting skill training programmes and achieving desired goals. These are:

1. Assessment

Identifying gaps that need to be addressed is the first step in skill training. This may involve individual assessments, performance evaluations, or broader analyses of organizational needs.

2. Curriculum

Designing a structured curriculum or learning plan that outlines the learning objectives, content, and activities necessary to achieve desired skill outcomes is crucial. The curriculum may include lectures, demonstrations, hands-on exercises, case studies, and assessments.

3. Instructional Methods

Selecting appropriate instructional methods and delivery formats based on the nature of the skills being taught and the preferences of the learners.

4. Trainers or Instructors

It is necessary to have qualified trainers or instructors with expertise in a particular subject matter to engage and motivate learners involved in the training process.

5. Practice and Application

Hands-on exercises, simulations, role-plays, and real-world projects allow participants to gain practical experience and reinforce their learning.

6. Feedback and Evaluation

Regular evaluations, quizzes, assessments, and performance reviews are essential for tracking learning outcomes and adjusting the training as needed.

Skill development is identified as the process of gaining, improving, or updating the skills, knowledge, and competences required to carry out tasks successfully in a variety of professional or vocational spaces. It includes programmes meant to hone soft skills like problem-solving, communication, and teamwork as well as technical skills like mastery of certain tools, technologies, or procedures. Skill development contributes to overall social and economic development by building human capital, reducing poverty, and promoting inclusive growth across communities and nations.

Skills management is a strategic approach to identifying, acquiring, developing, and utilizing the skills and competencies of individuals within an organization. It includes mapping and tracking skills that are present in the workforce, coordinating those skills with organizational objectives, and putting procedures in place to optimize these skills. The strategy of skills management is to maximize the pool of available talent to make sure that the organization has the right people in the right positions at the right time. Skills management is critical for building a talented and resilient workforce that can drive organizational success.

2.10. Skill and Competence: Conceptual Understanding

Competency in the American sense complements competence as used in the UK Occupational Standards. Synder and Ebeling (1992) refer to competence in a functional sense, but use competencies to refer traits in the plural. While some authors consistently use competency when referring to occupational competence (Smith, 1993), others treat the two as synonymous (Brown, 1993). Hartle (1995) argues that competency as a characteristic of an individual shown to drive superior job performance includes both visible competencies of knowledge and skills and underlying elements of competencies like traits and motives.

Elkin (1990) links competence with micro-level job performance and competency with higher management attributes and in defining managerial competencies for the future. Being competent refers to meeting the job demand, while having competencies refers to possessing the necessary attributes to perform competently.

In brief, it can be argued that competency refers to a group of characteristics, inherent or acquired, which indicate an individual's ability to acquire skills in a given sphere. Competence is related to the different aspects of the job which an individual can perform, while competency refers to a person's behaviour and underpinning competent performance.

Meta competence is concerned with an individual's knowledge of his own intellectual strengths and weaknesses, how to apply skills and knowledge in various task situations and how to acquire missing competencies. These include skills in planning, initiating, monitoring and evaluating one's own cognitive processes; knowledge about problem solving; skills in using effective cognitive aids and tools such as graphics and analogies.

Competency = Skill + Ability

Skill is learned, Ability is not learned, it is innate. While ability is the mental or physical power to do something based on the innate characteristic of an individual, skill stresses on learning acquired or developed through experience or practice.

Competence is broken down into three analytically distinct components. They are:

(a) **Conceptual competence-** abstract knowledge about an entire domain, ability to integrate information and make judgments using a number of relevant factors, implicit understanding of general principles of the domain.

(b) **Procedural competence-** procedure and skill needed to apply conceptual competence in concrete situation, understanding general principles of action.

(c) **Performance/utilization competence-** competence required to assess a problem and select a suitable strategy for its solution, understanding of relations between features of a task setting and requirements of performance (Sophian, 1997).

Cheetham and Chivers (1998) developed a holistic model of professional competence, comprising five sets of inter-connected competences and competencies. Their holistic competence framework comprises five dimensions:

(a) cognitive competence, including theory and concepts, as well as informal tacit knowledge gained experientially. Knowledge (know-that) underpinned by understanding (know-why), is distinguished from competence;

(b) functional competences (skills or know-how), those things that a person who works in a given occupational area should be able to do (and) able to demonstrate;

(c) personal competency (behavioural competencies, know-how to behave), defined as a 'relatively enduring characteristic of a person causally related to effective or superior performance in a job' ;

(d) ethical competencies, defined as 'the possession of appropriate personal and professional values and the ability to make sound judgements based upon these in work-related situations';

e) meta-competencies, concerned with the ability to cope with uncertainty, as well as with learning and reflection.

There is some confusion and debate concerning the concept of competence that makes it difficult to evolve a coherent theory, or to arrive at a definition capable of accommodating the different shades of meaning with which the term is used. Competency-based behavioural anchors are defined as performance capabilities needed to demonstrate knowledge, skill and ability acquisition. White (1959) is credited with having introduced the term competence to describe those personality characteristics associated with superior performance and high motivation.

McClelland (1973) followed this approach and developed tests to predict competence as opposed to intelligence, but subsequently he also described this characteristic underlying superior performance as 'competence'.

A distinction can be made between objective competence (performance and potential performance measured with standard tests) and subjective competence (assessment of abilities and skills needed to master tasks and solve problems relevant to performance). An example for objective competence is time taken to complete a given task, whereas the ability to work as a team is subjective competence.

Subjective competence is further divided into three components-

- Heuristic competence: generalized expectancy system concerning the effectiveness of one's abilities across different situations, eg: performance under different work environments.
- Epistemological competence: beliefs and confidence that one possesses, and domain specific skills and knowledge to master tasks and problems within a specific content domain, eg: mastery of IT skills.
- Actualized competence: momentary subjective self-confidence that one possesses, and the abilities, knowledge and skills believed necessary for success in a concrete learning or performance situation, eg: application of skills in problem solving.

Key competencies are content-independent, applicable and effective across different institutional settings, occupations and tasks. These typically include basic competencies such as literacy, numeracy, general education and methodological competencies like problem solving, IT skills, communication skills including writing and presentation and judgment competencies such as critical thinking.

2. 3 DeSeCo

The Organisation for Economic Co-operation and Development's (OECD) Definition and Selection of Competencies (DeSeCo) Project's conceptual framework for key competencies, classifies such competencies in three broad categories:

- Individuals need to be able to use a wide range of tools for interacting effectively with the environment: both physical ones such as IT and socio-cultural areas such as the use of language
- In an increasing interdependent world, individuals need to be able to engage with others
- Individuals need to be able to take responsibility for managing their own lives, situate their lives in the broader social context and act autonomously.

Source: https://www.oecd.org>edu>skill

The Mayer Committee Report (1992) marked the beginning of the establishment of what is referred to as Key Competencies in Australian education. It was formed to advise the Australian Education Council on employment related key competencies for post compulsory education and training. The Committee established the guiding principles to shape the conceptualization of generic skills into the Key Competencies which would eventually find their way into Training Packages.

The Key Competencies are described as:

- Essential to preparation for employment.
- Generic to the kinds of work and work organization emerging in the range of occupations at entry levels within industry.
- Equipping individuals to participate effectively in a wide range of social settings, including workplaces and adult life.
- Involving the application of knowledge and skill.
- Being able to be learned.
- Being amenable to credible assessment.

Key competencies heralded a new era in preparing learners for the reality of the workplace. However, it became increasingly apparent that they needed a) considerable review or restructuring to meet the needs of the late twentieth century workforce and b) much greater emphasis in the training system. Thus, the concept of Employable Skills has evolved.

2.11. Skill as a component of Behaviour

Though Bloom's taxonomy was primarily created for academic education to highlight the three components of behaviour, it is relevant for all types of learning. Bloom's taxonomy model covers three parts or 'overlapping domains'. They are:

(a) Cognitive domain (intellectual capability) - knowledge/ thinking

(b) Affective domain (feelings and emotions) - attitude/feeling

(c) Psychomotor domain (manual and physical actions) – skills/action

The learner should benefit from the development of knowledge and intellect (cognitive domain), attitude and beliefs (affective domain) and the ability to put physical and body skills into effect, in terms of action (psychomotor domain).

Cognitive **Knowledge**	**Affective** **Attitude**	**Psychomotor** **Skills**
1. Recall data	1. Receive (Awareness)	1. Imitation (Copy)
2. Understand	2. Respond (React)	2. Manipulation (Follow instructions)
3. Apply (Use)	3. Value (Understand &Act)	3. Develop precision
4. Analyze (Structure)	4. Organize personal value system	4. Articulation (Combine/ Integrate)
5. Synthesize (Create)	5. Internalize value system (Adopt behaviour)	5. Naturalization (Automate)
6. Evaluate (assess)		

Ref: Bloom's Taxonomy Domains: Cognitive Domain- Affective Domain-Psychomotor Domain

A. Cognitive Domain

The cognitive domain involves the development of our mental skills and the acquisition of knowledge. The six categories under this domain are:

1. **Knowledge:** the ability to recall data and/or information.

   Example: A child recites the English alphabet.

2. **Comprehension:** the ability to understand the meaning of what is known.

   Example: A teacher explains a theory in his own words.

3. **Application:** the ability to utilize an abstraction or to use knowledge in a new situation.

   Example: A nurse intern applies what she learned in her Psychology class when she talks to patients.

4. **Analysis:** the ability to differentiate facts and opinions.

   Example: A lawyer winning over a case after recognizing logical fallacies in the reasoning of the offender.

5. **Synthesis:** the ability to integrate different elements or concepts in order to form a sound pattern or structure so a new meaning can be established.

   Example: A therapist combines yoga, bio-feedback and support group therapy in creating a care plan for his patient.

6. **Evaluation:** the ability to come up with judgments about the importance of concepts.

Example: A businessman selects the most efficient and successful way of selling products.

B. Affective Domain

The affective domain involves our feelings, emotions, and attitudes. This domain is categorized into five subdomains, which include:

1. **Receiving Phenomena:** the awareness of feelings and emotions as well as the ability to utilize selected attention.

   Example: Listening attentively to a client.

2. **Responding to Phenomena:** active participation of the learner.

   Example: Participating keenly in a group discussion.

3. **Valuing:** the ability to see the worth of something and express it.

   Example: A trade union leader shares ideas on wage hike for labourers.

4. **Organization:** ability to prioritize a value over another and create a unique value system.

   Example: An employee spends more time in learning a task, rather than spending time in entertainment activities.

5. **Characterization:** the ability to internalize values and let them control the person`s behaviour.

   Example: A person takes up a job not considering the salary, but his passion for it.

C. Psychomotor Domain

The psychomotor domain is comprised of utilizing motor skills and coordinating them. The seven categories under this include:

1. **Perception:** the ability to apply sensory information to motor activity.

   Example: A cook adjusts the heat of stove to achieve the right temperature of the dish.

2. **Set:** the readiness to act.

   Example: An obese person displays motivation in performing planned exercise.

3. **Guided Response:** the ability to imitate a displayed behavior or to utilize trial and error.

   Example: A person follows the procedure manual in operating a machine.

4. **Mechanism:** the ability to convert learned responses into habitual actions with proficiency and confidence.

   Example: A mother was able to cook a delicious meal for her child after practicing how to cook it.

5. **Complex Overt Response:** the ability to skilfully perform complex patterns of actions.

   Example: Preparing a report using AI.

6. **Adaptation:** the ability to modify learned skills to meet special events.

   Example: A designer uses plastic bottles to create an art form.

7. **Origination:** creating new movement patterns for a specific situation.

   Example: A choreographer creates a new dance routine.

The psychomotor domain explains skills development relating to manual tasks and physical movement. It also concerns and covers modern day business and social skills such as communication skills and IT skills. Thus 'motor skills' extend beyond the original traditionally imagined manual and physical skills. It is always necessary to use this domain, even if one thinks that his/her environment is covered adequately by the cognitive and affective domains. Whatever be the situation, it is likely that the psychomotor domain is significant.

The Dave version of the Psychomotor domain is considered the most relevant and helpful for work and life related development, more precisely skill development.

Dave's psychomotor domain taxonomy

Sl. No	Category/ Level	Behaviour descriptions	Activity/ Measurement	Key words
	Imitation	Copy action, Observe & replicate	Watch & repeat action	Copy, follow, repeat
	Manipulation	Reproduce from memory	Carry out task based on instruction	Recreate, Build. Execute, Implement
	Precision	Execute skill independent of help	Perform with expertise and quality	Demonstrate, Calibrate, Control
	Articulation	Adapt and integrate expertise	Relate & combine associated activities	Construct, Solve, Integrate, Modify
	Naturalization	Automated unconscious mastery of activity	Define aim and strategy for use of activities	Design, Specify, Invent

2.12. Skill gap

Skill gap can be explained as 'redefining the relationship between education, industry and business.' In simple terms, a skill gap can be defined as the difference between the skills needed for a job versus those skills possessed by a prospective worker.

There are two types of labour force entering the labour market due to their poor economic conditions. One is the educated labour force who are not able to find jobs matching their qualification due to lack of technical or soft skills. This is the reason for the high rate of unemployment among the educated youth. To reduce the skill gap among the educated, there is need for better quality education, knowledge of English language, on the job training as well as possession of appropriate skill sets.

Those countries that have succeeded in linking skill development to gains in productivity, employment and development have targeted skill development policy towards three main objectives:

1. Matching supply to current demand for skills,
2. Helping workers and enterprises adjust to change; and
3. Building and sustaining competencies for future labour market needs.

The **first** objective is about the relevance and quality of training. Matching the provision of skills with labour market demand requires labour market information systems to generate, analyse and disseminate reliable sectoral and occupational information, and institutions that connect employers with training providers. It is also about equality of opportunity in access to education, training, employment services and employment, in order that the demand for training from all sectors of society is met.

The **second** objective is about easing the movement of workers and enterprises from declining or low-productivity activities and sectors into expanding and higher productivity activities and sectors. Learning new skills, upgrading existing ones and lifelong learning can help workers to maintain their employability and enterprises to adapt and remain competitive in the field.

The **third** objective calls for a long-term perspective, anticipating the skills that will be needed in the future and engendering a virtuous circle in which more and better education and training fuels innovation, investment, technological change, economic diversification and competitiveness, and thus job growth.

Skill Gap Analysis

One of the ways organisations can stay ahead of the curve is to conduct a skill gap analysis to ensure that their workforce stays productive, improves performance and stays competitive in a rapidly changing world. It is a tool that an organisation can use to improve performance by identifying the gaps and taking action to bridge it to stay competitive in the long run.

The procedure involves the following steps:

1. Identify current skill availability:

Assess the present skills and skill levels available by conducting a skills mapping exercise. Organisations can often benefit if they ensure that skill mapping is an inclusive process. Besides the organisation mapping skills (top down), employees can be asked to map their own skills (bottom up) where they list their strengths, record their expertise levels as well as list the skills they enjoy using at work providing them energy and satisfaction.

2. Identify the skills needed in the future:

- Ask what are the organisation's goals for the next five years. For instance, whether there is plan to pivot the business strategy or model anytime in the next five years. If so, list the skills needed to fulfil those new goals.
- Do the current employees have the skills needed for the anticipated organisational changes or market disruptions? If no, what are the skill sets needed?

3. Fix a skills gap hierarchy:

Now, there are two lists of skills: a) The skills the organisation currently has, and, b) the skills for employees needed in the future. The difference between the two lists is the skills gap.

4. Design a solution:

Upskilling and reskilling through training and development is the most popular solution to bridging skills gaps.

2.4 World Bank study recommends skill-based training to enhance jobs in multiple sectors

The Services sector jobs and agriculture income enhancement present large opportunities for school-based skilling, according to "Jobs at Your Doorsteps," a skill gap study prepared by the World Bank based on availability of skill education and jobs in Himachal Pradesh, Kerala, Madhya Pradesh, Maharashtra, Odisha, and Rajasthan. The study also found that 14 sectors and corresponding trades can cater to majority of high demand jobs for school to work. The report was released in New Delhi in November 2024 jointly by Union Education Minister and Union Labour Minister.

The study examined key questions such as how can the skill education be scaled and strengthened and which all skills should be offered considering local economic needs. It also dealt on the ways to strengthen outcomes of skill education such as jobs, self- employment, relevant future higher education or further skilling pathways.

Source: The Hindu, Nov 23, 2024

2.13. Skill Benchmarking

While skill gap analysis identifies the difference between current employee skills and those needed to meet business goals, skill benchmarking evaluates employee skills against industry standards or best practices.

Skill benchmarking is similar to measuring how good someone is in performing some activity by comparing them to a standard. It is important today because it helps people and companies understand where they stand in their work compared to others. Skill benchmarking is like a map that helps one to know where one is and where he/she needs to go to be a better performer.

The different steps in benchmarking are:

1. Define Skill Requirements: Clearly define the skills and competencies required for a particular role or task and establish benchmarks or standards against which skills will be measured. This could involve technical skills, soft skills, knowledge, and other abilities necessary for successful job performance.
2. Assessment: Assess the skills of individuals or the organization as a whole using various methods such as tests, interviews, performance evaluations, simulations, or 360-degree feedback. This assessment should be aligned with the defined skill requirements and benchmarks.
3. Compare Skills to Benchmarks: Compare the assessed skills against the established benchmarks to identify any gaps or differences. This comparison highlights areas where individuals or the organization fall short of the desired skill levels.
4. Identify Skill Gaps: Analyse the differences between the assessed skills and the benchmarks to identify specific skill gaps. These skill gaps represent areas where individuals or the organization need to improve to meet the desired skill standards.
5. Develop Action Plans: Develop action plans to address the identified skill gaps. This may involve providing targeted training, mentoring, coaching, on-the-job learning opportunities, or other developmental activities aimed at closing the gaps and enhancing skills.

Learning Experience Platforms (LXPs) help in skill benchmarking by providing personalized learning experiences tailored to individual skill levels and job roles. Many LXPs offer built-in skill assessment tools and features that enable users to evaluate their current skill levels in various areas. LXPs can integrate with external skill benchmarks, such as industry certifications, standards, or frameworks. This integration allows organizations to align their learning initiatives with recognized benchmarks and ensure that learners are acquiring skills that are relevant and valued in their respective industries.

2.14. Skill Value Chain

The internal value chain of a private training provider is most simply defined as the activities that a trainee goes through in a skills programme for employment. It reflects how various stakeholders provide input at different stages of skills design and delivery. Using Porter's value-chain approach, these practices are defined as below:

- "Primary activities" for skills businesses include: (1) sourcing, (2) development of training standards and content, (3) demand-oriented training delivery, (4) assessment and certification, (5) placement and (6) post-placement support.
- "Secondary" or "support activities" include: (7) partnership, (8) financing, (9) technology and (10) monitoring and evaluation.

Sourcing is defined as "identification, assessment and engagement of skilled worker candidates through proactive recruiting techniques as well as mapping and linking with the available job opportunities." Sourcing consists of sourcing of trainees, sourcing of jobs and pre-training assessments.

Training standards and content refers to the development of the training courses, aligning curriculum with industry needs, offering market-relevant skills and life/soft skills, synchronizing the training offer with the occupational standards. It relates to regulations and practices with respect to the design of training programmes in line with the qualification frameworks, covering components such as curriculum development and trainer qualifications. National qualification frameworks help to overcome the skills mismatch challenge and to define the qualifications and assure that the training quality conforms to them.

Demand-oriented training delivery: The training stage consists of the methodologies, techniques and tools in the training programmes in addition to the on-the-job trainings and apprenticeships that highlight technical competencies. Training delivery should combine practical and theoretical learning. Innovations emerge when the programmes are complemented with multimedia and online technologies.

Assessment and certification are both the appraisal of an individual's knowledge and abilities vis-à-vis a selected set of criteria, and its validation through a standardized evaluation process. In the assessment and certification stage, it is important to align trainings with national occupational standards and national qualification frameworks.

Placement is the provision of wage employment opportunities to trainees after the completion of skills training programmes. For the placement stage, employment linkages with the private sector are required to ensure that training leads to employment. The absence of strong placement linkages with industry is a major area which requires improvement in India's skills development system.

Post-placement support is the totality of services offered to employers and employees for the continuation of the employment relationship. This includes tracking and handholding process that follows the placement of a trainee in a job.

It is essential to sustain the results of training, and trainees may require assistance and guidance to adapt to the work life and to succeed in their new jobs.

Partnership: For business models to successfully serve the needs of all social groups, partnerships with different stakeholders are instrumental. The private sector is a vital partner in the design and delivery of skills. Placement partnerships with businesses is another key type of cooperation in the skill sphere.

Another support activity, **financing,** provides an analysis of the different sources of funding tapped for the trainings, such as public funds, CSR funds and learner-paid financing, as well as key strategies to achieve financial sustainability.

As one of the support activities, **technology** refers to the innovative use of information and communications technology to achieve scale and impact. Technology is often used by the skills providers to scale up their business models in India and to promote learning and sharing of resources, knowledge and good practices as well as for quality control and achievement of scale.

2.5 K-YAN of IL & FS

Technology has been used extensively by skills providers. IL&FS Skills uses K-Yan, a patented multimedia device and community computer that enables delivery of interactive training with minimal infrastructural requirements, especially in spoke centres. K-Yan has the combined aspects of many multimedia devices, along with internet connection. The organization standardizes its training offer through the use of K-Yan, which helps overcome constraints such as lack of skilled trainers and absence of classroom infrastructure.

- Single cable device with computer, portable interactivity, projector and internet connectivity
- Converts any wall into an interactive smart board
- Comes loaded with multimedia content for 360 degree learning

Source: https://old.ilfsindia.com>stories

Monitoring and evaluation tracks results along the skills value chain, including challenges and business success. Monitoring and evaluation practices are critical to assess which skills training modalities produce the desired results, which in turn indicates which should be considered for replication and scale.

LabourNet has developed a comprehensive impact assessment framework that takes into account training outcomes (the "breadth of impact"), such as number of people trained and number of trainees certified, as well as broader socioeconomic results (the depth of impact), such as wage/income increase, enhanced quality of employment (e.g; reduced degree of attrition and absenteeism) and productivity increase.

References

Australian Education Council, 1992 Mayer Committee Report, Australian Education Council and Ministers of Vocational Education, Employment Training, 1992

Brown, R B 1993 Meta competence: a recipe for reframing the competence debate, Personnel Review, Vol 22, No. 6, p 25-36

Centre for Instructional Technology and Training 1956 Bloom's Taxonomy, https// citt.ufl.edu>resources>blooms-taxonomy

Cheetham, G., Chivers, G 1996 Towards a holistic model of professional competence, Journal of European Industrial Training, Vol 20, No.5, p 20-30

Coursera, 2024 What is microlearning? Definition, Use cases and More, https:// www.coursera.org>enterprise>articles> what- is

DFID 2007 Annual Report, https://www.gov uk>government> publications>department for international development-annual report

Elkin, G 1990 Competency based human resource development, Industrial and Commercial Training, Vol 22, No.4, p 20-25

Hartle, F 1995 How to re-engineer your performance management process, London: Kogan Page

Hillage, J and Pollard, E 1998 Employability: Developing a Framework for Policy Analysis, Department of Education and Employment, London

McClelland, 1973 Testing for Competence rather than for 'Intelligence', https:// www.scirp.org>reference>reference papers

NITI Aayog 2015 Skill development and Productivity of the Workforce,

NSSO 2001 Concepts and Definitions Used in NSS, Ministry of Statistics and Programme Implementation, Government of India, https://www.mospi. gov.in>concepts_golden

Planning Commission 2007 Report of the Task Force on Skill Development, Planning Commission, GoI, May 2007

Planning Commission 2020 India Vision 2020, https://library.niti.gov.in>report-india- vision 2020

Pear, T H 1927 Skill, Journal of Personnel Research, Vol 5, p 478-489

Proctor R W., Dutta, A 1995 Skill acquisition and human performance, Sage, London

Smith, B 1993 Building managers from the inside out: competency- based action learning, Journal of Management Development, Vol 12, No.1, p 43-48

Sophian, S 1997 Beyond Competence: the significance of performance for conceptual development, Vol 12, p 281-303

Snyder, A., Ebleling, HW 1992 Targeting a company's real competencies, Journal of Business Strategy, Vol 13, No.6, p 26-32

Utica University Bloom's Taxonomy of Measurable Verbs, https://www.utica.edu>Bloom's Taxonomy

Welford, A T 1968 Fundamentals of Skill, Mathuen, London

Welford, A T 1976 Skilled performance: Perceptual and motor skills, Glenview, Scott

White, R W 1959 Motivation reconsidered: The concept of competence, Psychological Review, 66 (5), 297-333

Winterton, J et al 2006 Typology of knowledge, skills and competencies: clarification of the concepts and prototype, Cedefop Reference Series 64, European Centre for the Development of Vocational Training, Luxembourg

CHAPTER 3

SKILL DEVELOPMENT: ISSUES AND CHALLENGES

Over the course of the 20th century, the mature economies of the world evolved from **industrial** economies to **knowledge** economies. Skills and knowledge are the driving forces of economic growth and social development of any country. As India moved progressively towards becoming a 'knowledge economy', there was need for focused efforts on advancement of relevant skills for the emerging economic environment. Whereas in old economy, skill development largely meant development of shop floor or manual skills, in the knowledge economy, the skill sets range from professional, conceptual, managerial, operational, behavioural to interpersonal skills and inter-domain skills.

Presently, we are at another watershed moment, transitioning to **human** economies, and the shift has profound implications for skill development. Economies get labelled according to the work people predominately engage in. The industrial economy replaced the agrarian economy when people left farms for factories and then the knowledge economy pulled them from factories to office buildings. Instead of leveraging their brawn, companies capitalized on their brains. In the place of hired hands, hired heads dominated.

3.1 Knowledge Economy

Nick O' Donovan's "Pursuing the Knowledge Economy: A Sympathetic History of High Skill, High Wage Hubris" is an interesting evaluation of the policy consensus of the 1990s and early 2000s concerning the opportunities afforded by digital technology and globalisation for a transition to better jobs in the western economies. It politely critiques the knowledge economy analysis in itself, essentially for unfounded optimism about the scope for new technologies to complement rather than substitute for labour and for ignoring the transition costs associated with globalisation. The book ends by linking the failure of the Knowledge Economy growth regime to post-2008 political polarisation, and hoping that the current period of disillusion and volatility will in fact prove an opportunity for a new, better "growth regime" to develop.

Source: https://www.agendapub.com

In the human economy, the most valuable workers are hired hearts. The know-how and analytic skills that made them indispensable in the knowledge economy no longer give them an advantage over increasingly intelligent machines. But still they bring to their work essential traits that cannot be and will not be programmed into software, like creativity, passion, character, and collaborative spirit, in other words, humanity.

In India, as per the Periodic Labour Force Survey (PLFS) Annual Report 2021-22 of National Sample Survey Office (NSSO), percentage of persons in the age group 15-59 years who received vocational / technical training was 19.5 percent, which includes 3.4 per cent who received formal vocational / technical training, indicating that only low number of young generation actually enter the world of work with any kind of formal vocational training. This proportion of trained youth is probably one of the lowest in the world. Switzerland, Denmark, Hongkong, South Korea, Netherlands and Singapore are some of the top performing countries in this regard.

The lack of awareness about benefits of vocational training and its perceived low aspirational value are among important reasons for low participation in vocational/technical training programmes.

As detailed in Chapter 1, for a country's economic development, as well as social growth, skill development plays a crucial role. Particularly for a country like India, there is a constant requirement for higher employment opportunities to match its massive population, with millions of freshers entering the job market every year. India is apparently at an advantageous position with its unique demographic dividend, compared to its foreign counterparts including the USA, UK and Japan. While for these countries, in the next 20 years, the labour force will decline by 4 per cent in industrialized world, India will have a much younger labour force with an increase by 32 percent. This trend will continue for the next 50 years, and will provide immense opportunities to better shape the country's future. However, to reap its benefits, India needs to equip its workforce with employable skills and knowledge so that they can contribute substantively to the economic growth of the country.

A review of the existing vocational training system in India reveals that skill acquisition takes place through two basic structural streams- a small formal one and a large informal one. The formal structure includes- i) higher technical education imparted through professional colleges, ii) vocational education in schools at the post-secondary stage, iii) technical training in specialized institutions, and iv) apprenticeship training.

The informal economy in India provides more than 90 per cent of the labour force. This leads to the largest number of people in informal settings being trained informally and their skills are considered to be low. Informal skills refer to "learning resulting from daily activities related to work, family or leisure". It is not organised or structured in terms of objectives, time or learning support.

Informal training is an unstructured and unofficial method of knowledge sharing, which can entail knowledge transfer between senior persons and their juniors. It can also entail a new hire learning from an experienced person (eg: masonry skills, tailoring skills, carpentry, etc).

3.1. Vocation and Career

In India, the concepts of vocation and career are deeply intertwined with societal norms, economic conditions, and individual aspirations. While historically, vocation often referred to traditional occupations passed down through generations, the modern landscape views a significant shift towards career-oriented pursuits influenced by globalization and technological advancements. In India, vocation often aligns with caste-based professions such as carpentry, farming, weaving, and pottery, where skills are passed down from one generation to another. These vocations are not merely means of livelihood but also integral to cultural identity and social cohesion.

With the advent of modern education and exposure to diverse career options, individuals are redefining their vocational pursuits beyond inherited professions. There is a growing emphasis on skill development, entrepreneurship and innovation, reshaping the landscape of vocation across the country.

Generally, vocation and career are used interchangeably. Vocational education might be classified as teaching procedural knowledge. This may be contrasted with declarative knowledge, as used in education in a usually broader scientific field, which might concentrate on theory and abstract conceptual knowledge, characteristic of tertiary education. In contrast to vocation, a career refers to a series of progressive roles or occupations pursued over a person's working life, often driven by personal goals, professional growth, and financial stability.

Vocational education can be at the secondary or post-secondary level and can interact with the apprenticeship system. Increasingly, vocational education can be recognised in terms of recognition of prior learning and partial academic credit towards tertiary education (e.g., at university level) as credit; however, it is rarely considered in its own form to be included under the traditional definition of a higher education.

Until the end of the twentieth century, vocational education focused on specific trades such as for example, an automobile mechanic or welder, and was therefore associated with the activities of lower social classes. As a consequence, it attracted a level of stigma. However, as the labour market becomes more specialized and economies demand higher levels of skill, governments and businesses are increasingly investing in the future of vocational education through publicly funded training organizations and subsidized apprenticeship or traineeship initiatives for businesses. At the post-secondary level vocational education is typically provided by an institute of technology, or by a local community college.

3.2 Vocation & Career-Challenges

Despite the evolving landscape, vocation and career in India face several challenges. The persistent influence of societal norms, gender biases, and regional disparities often restrict individuals' choices and access to opportunities. Moreover, the mismatch between educational curricula and industry requirements poses hurdles for skill development and employability.

However, amidst these challenges lie significant opportunities for growth and innovation. Government initiatives such as Skill India, Start-up India, and Digital India aim to foster entrepreneurship, enhance employability, and bridge the gap between education and industry.

3.2. Changing scenario of skill development in India

The vocational training and skill development landscape has been evolving in India from the post-independence era. The initial focus was on establishing a formal technical vocational education training or TVET sector, with dedicated technical and vocational institutions mainly designed for the manufacturing and engineering trades. This era was marked by establishing Industrial Training Institutes (ITIs) in 1970s.

With the advent of economic liberalisation in India during the 1990s, new sectors grew rapidly, including the IT industry and a growing service sector. This led to a paradigm shift and the need to rapidly expand the capacity of skills training sector was recognised. A significant effort was made to bring more private sector organisations into the skilling ecosystem.

Over the years, the skilling landscape in India has strengthened and there are various implementation models for activities within the skill value chain. The skill providers – public, private or public private partnerships – are innovating in different ways to mobilise the right candidates for skill courses.

The brick-and-mortar type of training centres evolved adjusting to the present needs and at present, we have mobile training centres, skill vans, blended learnings, learning at the workplace and many other such models. Skilling has become aspirational and it is converging with formal education pathways through establishment of Skill Universities now providing degree programmes.

In many developing countries like India, where the government plays the dual role of policy maker and regulator, skills training is largely government-driven and constant efforts are made to increase private sector participation to perform the role of a training provider (TP) and for successful absorption of trained youth.

Despite the substantial efforts made in the last few years to augment the TVET ecosystem, the sector still faces many challenges in strategically channelising the young population towards vocational education to reap the massive demographic dividend in the country.

Some of the models deployed in skill development are:

1. Counselling and Career Planning (Community Model): This model involves reaching out to the communities and partnering with them to organize awareness events, mobilization fairs or setting up dedicated mobilization centres.

2. Use of Technology in Counselling and Career Planning: In addition to the manual process of counselling, the TPs, post mobilization, have been increasingly relying upon technology, such as e-psychometric tests, to better gauge the aspiration level of the trainees and suggest a suitable career path. These online platforms blend artificial intelligence and machine learning solutions with strategic human interventions to improve the decision making for the end user.

3. Various training delivery models have evolved in the last decade such as conventional classroom learning, skills on wheels and blended online learning programmes, and Massive Open Online Courses (MOOCs).

    a. Classroom learning: It is a conventional setup consisting of trainers, students, paper based vocational content and curriculum. By industry standards, one classroom consists of 30-40 students per session, lasting for one and a half to two hours.

    b. Skill on Wheels: Short-term skills recognition model (especially for prior learning or upskilling) is adopted by a few states and TPs in India. In this, a hi-tech vehicle moves around geographies in order to reach out to the rural youth. The vehicle has pre-planned stops wherein it takes a halt for a period of four to five days and sets up a mini-facility to upskill the local youth. The vehicle is equipped with Pedagogic tools and equipment that can be used for training readily during these stops.

    c. Massive Open Online Courses (MOOCs): This model uses the power of technology to provide just-in time trainings at speed and scale, and offers vocational education for unlimited participation at a low cost. The typical MOOC features include open access, free enrolment, massive scale, instantaneous assessments and results, and host of industry benchmarked courses.

    d. Online training methodology consists broadly of four components – video lectures, downloadable content, question and answer forums, and help from e-trainers to clarify doubts. Some platforms have premium features such as online assessments, online score card and recommendations on bridge courses etc.

4. Centre of Excellence (CoE). They are places where the highest standards are maintained in respect to skill development in a specific sector(s). CoEs for skilling include structures, entities that leverage shared resources from the private sector, government and public sector. By bringing the various stakeholders together, CoEs can support adoption of best practices, superior leadership, research and high quality training by focusing on a specific sector. Contributions can be made by various stakeholders in setting up these CoEs.

3.3. Skills business ecosystem in India

Government and private sector partners in India have intensified their collaboration to create a conducive environment for business, which would lead to inclusive growth to benefit the poor and underserved. The government provides direction to the private sector, which operates according to a broader vision in skills development.

To govern the skills business ecosystem in the country, India created many different policy measures, notably the skill development and entrepreneurship policy and NSDC. These policies were **ecosystem- level interventions** as the Government needed to fundamentally change the skilling industry to meet the enormous challenge it had set itself of skilling millions.

NSDC and the National Skill Development Agency have identified five key elements that the business ecosystem should possess.

- **Cradle to career approach:** Skills development should be viewed as a continuous process to overcome the barrier of the knowledge and skills of a person becoming obsolete. This should be accompanied with a lifelong learning perspective, offering various livelihood opportunities to individuals while responding to the human resource needs of the industry.
- **Digitized marketplace:** Digital technologies enable efforts to achieve the large-scale training targets of India, tailored to also meet the skills needs of hard-to-reach segments of the population. Technology should be leveraged to bring together youth seeking job opportunities, training providers (institutions) and prospective employers.
- **Standardization and quality:** Standardization and quality in the market will help align the expectations of the Indian industry with the outputs of training programmes. National occupational standards and the set of standards associated with a job role (qualification packs) should be in sync with global standards, enabling youth to tap job opportunities not only at the local and national levels but also at the global level.
- **Vocational education by choice:** An important challenge for skills development is the low level of attractiveness of vocational skilling programmes and jobs in India. While keeping demand for skills is crucial for the existence of the ecosystem, vocational education and training

should be by choice. It should become an aspirational, desired career path for the youth, while the importance of vocational skills should be acknowledged and industry should accept wages linked to skills and competencies.

- **Talent capital of the world:** In order for India to achieve its goal of becoming the "talent capital of the world", it will need to lower the costs of acquiring new talent and reaching scale. A vibrant skills ecosystem will lower the costs and increase the scale of skills delivery. This is essential if the country is to harness its demographic dividend, utilizing its vast working-age population in a productive manner.

3.4. Platform labour and platform economy

Digitalisation/platformization of work has given rise to a new classification of labour, platform labour, different from the traditional dichotomy of formal and informal labour. Platform labour has emerged and grown over the last decade, thanks to the omnipresence of smartphones and mobile data. Since the entry barriers to such work are low, platforms hold enormous potential to unlock millions of jobs in India.

The contemporary labour market has witnessed the emergence of two new categories of work- gig and platform work. Gig and platform work has emerged as a key employment generator globally and also in India. NITI Aayog estimates that the sector had 7.7 million workers in 2020, and the number is set to triple to 23.5 million by 2029 – 30. The sector has also led to drastically shifting work arrangement practices in the last few years, giving rise to new issues relating to employment, to which Employers and Business Member Organizations (EBMOs) globally have traditionally responded within a regulated framework.

Currently, e-commerce platforms and online aggregators of goods and services are registered first under the Companies Act, 2013 and are regulated as technology companies under the IT Act, 2000. Empowered by the Consumer Protection Act, 2019, the Ministry of Consumer Affairs notified the Consumer Protection (E-commerce) Rules in 2020.

The platform economy encompasses economic and social activities facilitated by digital platforms. These platforms such as Amazon, Airbnb, Uber, Microsoft and Google serve as intermediaries between various groups of users, enabling interactions, transactions, collaboration, and innovation.

These digital platforms are diverse in function and structure. Google and Facebook are digital platforms that offer search and social media, but they also provide an infrastructure on which other platforms are built. Amazon is a marketplace, as are Etsy and eBay. Amazon Web Services provides infrastructure and tools with which others can build yet more platforms. Airbnb and Uber use these newly available cloud tools to force deep changes in a variety of incumbent businesses. Together they are provoking reorganization of a wide variety of markets, work arrangements, and ultimately value creation and capture.

This digitally based new economy has been given a variety of names derived from some of its perceived attributes. How we label this transformation matters because the labels influence how we study, use, and regulate these digital platforms. Its boosters have called it the Creative Economy or the Sharing Economy, whereas those less convinced of its benefits have dubbed it the Gig Economy, the Precariat, or the 1099 Economy, focusing on its impact on workers and how they are compensated.

3.5. Platform led skilling as a growth model

In many instances, the platform companies which provide earning opportunities for those associated with them take considerable steps in skilling these workers as well. This is to assure uniform standards in quality of service and to gain repeat business through the app itself, while also achieving upskilling of workers in an otherwise highly unorganised labour economy. Some of the initiatives in this direction are helmed by platform companies tied up with the National Skill Development Corporation (NSDC) to upskill blue-collar workers at scale. This is crucial given the smart city aspirations of many urban centres in India, and such Recognition of Prior Learning (RPL) and certification of upskilling signals the potential for workers to build the smart cities of tomorrow (Osborn, 2020). This tie-up is set to benefit various sectors such as beauty and wellness, plumbing, carpentry and other home improvement works, as well as servicing and repair, by creating a technology and skill mediated career progression.

India's huge informal labour market also stands to benefit from a restructuring due to the platformization of skilling and jobs. Several initiatives by e-commerce and platform companies aim at gender, disability and other marginalisation-inclusivity of such a diverse workforce while simultaneously bringing more informal workers under the fold of an economy of jobs which offers better incomes and experience of work. By combining income generation, gainful self-employment and new forms of social protection, technology platforms have the potential to create a better working environment for all.

NITI Aayog (2022) had published a report on the gig platform economy in India. According to the report, Gig workers, those engaged in livelihoods outside the traditional employer-employee arrangement, can be broadly classified into platform and non platform-based workers. Platform workers are those whose work is based on online software apps or digital platforms. While, non-platform gig workers are generally casual wage workers and own-account workers in the conventional sectors, working part-time or full time.

3.6. Major issues related to Skill Development

In contemporary analyses of systems of skill development in developing countries, the following issues are flagged:

- Importance of developing effective skill development systems in the interest of enhancing productivity and sustaining high rates of growth.
- Changing skill requirements of economies witnessing fast changes in the nature of economic activity and work.
- Need to match skills to markets, financial sustainability of different skill formation systems and the role of different agents in delivering the skill training efficiently and effectively.

The important issues related to skill development are:

- **Inadequate scale, limited capacity:** The existing infrastructure, both physical and human, is grossly inadequate considering the projected demand for skilled labour. While there is a need to create additional capacity in existing institutes, there is also a need to create an adequate infrastructure even in small towns and villages. In terms of faculty also, the training infrastructure is inadequate.
- **Access and completion:** Across the world, investments in education and skills development, from preschool through post-secondary education to vocational training, have high returns. Still, provision of equitable access is a challenge in many low-income and middle-income countries. Furthermore, many students who manage to enrol in education or training programmes do not complete their studies and miss out during the course of education/training, which can dramatically reduce the return on the educational investments in terms of lifetime earning potential.
- **Awareness, mindset and perception issues:** Skill development in India is way below the requirements due to lack of awareness on the type of courses as well as information on the ensuing career prospects. More importantly, there is limited acceptance of skill development courses as a viable alternative to formal education. Skilling is often viewed as the last resort meant for those who have not been able to progress in the formal academic system. This is partly due to lack of integration between the two options and also due to rising aspirations for white collar jobs which necessitate higher academic qualifications. Skill development is also often associated with blue collar jobs, which is largely perceived to be of low dignity and provides low wages/salaries.
- **Quality concerns:** Many young people attend schools without acquiring basic literacy skills and hence they are unable to compete in the job market. For those who access technical and vocational training at secondary and post-secondary levels, returns can vary substantially by specialization and institution. In particular, technical and vocational training (TVET) systems in many countries face challenges related to quality assurance, resulting in perceptions of the vocational track being a second-best option compared to general secondary or tertiary education.

There is a serious mismatch between the industry's requirements and the skills imparted in educational and training institutes, especially for the mid-level skills requiring some expertise on handling of machinery. To tackle this problem, considerable improvement of the quality of training is needed.

The availability of good quality trainers is also a key concern. The quality of trainers is affected due to limited efforts towards re-training and skill improvement of trainers. There is a lack of focus on development of trainers with a clear career path which can make this an aspirational career choice and can ensure regular adequate supply of good quality trainers in every sector.

- **Relevance:** Technical and vocational education and training, which can last anywhere from six months to three years, can give young people, especially women, the skills to compete for better paying jobs. However, more needs to be done in terms of engaging local employers to ensure that the curriculum and delivery of these programmes responds to labour market needs.
- **Efficiency:** The challenges related to governance, financing, and quality assurance also impact the efficiency of skills development programmes. The resulting high costs can limit opportunities for disadvantaged youth and adults to access these programs.
- **Cost concerns:** Skill development initiatives in India continue to be largely dependent upon the government funds or public-private ventures. Owing to high capital requirements and low return on investments, skill development is often viewed as a non-scalable model and remains under-invested. However, the fee-based model also faces challenges as prospective students are often not willing or unable to pay high fees for training. Even the willingness of commercial banks to lend for skill development activities is very low as educational loans are perceived as high risk products due to uncertainty with respect to future employment.
- **Mobility concerns:** In India, educational qualification is generally preferred over vocational training as former is associated with better employment opportunities, in terms of pay as well as quality of work. Additionally, there is limited mobility between formal education and vocational training in India due to lack of equivalent recognition for the latter. A student enrolled in vocational training often cannot migrate to institutes of higher education due to eligibility restrictions.

3.7. Challenges to skill development

Some of the challenges within India's skilling and entrepreneurship landscape are:

1. Public Perception/Mindset: Skilling is often perceived as a last resort for those who have not progressed within the formal academic system

2. Complexity in Assessment and Certification: Multiple assessment and certification systems lead to inconsistent outcomes and confusion among employers
3. Demand-Supply Mismatch: Discrepancies exist between skill supply and demand at both sectoral and spatial levels
4. Limited Mobility: Transitioning between skill-based and higher education programmes is often difficult under Indian conditions
5. Apprenticeship coverage: Apprenticeship programmes have low coverage and acceptance in the Indian context.
6. Obsolete Curricula: Skill curricula are often narrow and outdated and do not match with the industry needs
7. Gender Disparities: The declining labour force participation rate among women affects skill development
8. Unorganized Sector Employment: Non-farm, unorganized sector jobs lack productivity incentives for skilling
9. Entrepreneurship Education Gap: Entrepreneurship is inadequately integrated into formal education
10. Startup Challenges: Lack of mentorship and limited access to finance hinder the success and sustainability of startup ventures.
11. Innovation Impetus: Innovation-driven entrepreneurship needs greater emphasis, which is very much lacking in the skill sector.
12. Wage Premium: Skilled workers often lack assured and acceptable wage premium

India's eventual success in skilling the millions will depend on formulating and disseminating innovative solutions to these challenges, as indicated below:

- Skills mismatch - First, skills supply does not always correspond to skills demand. As a result, the skills supplied might not link with the needs of the market.
- Absence of capacity and quality- The number of people in India undertaking vocational skills courses in both private and government ITIs is approximately 2.19 million, a negligible number compared with the 460 million individuals in the labour force. The quality of applied trainings at ITIs is far below the required levels, which is also due to the low level of connection between the industry and education and training institutions.
- Recognition of informal acquisition of skills (prior learning)- There is a pressing need for mechanisms to analyse skills acquired informally and certify them. However, it is worth noting that RPL has been initiated under the PMKVY scheme.

- Apprenticeship could be leveraged to facilitate skills acquisition as it fosters both "learning by doing" and "learning by earning". However, the country only has around 341,000 apprentices, which is a very low number considering that its labour force exceeds 450 million.
- The competencies and knowledge of the instructors cannot keep up with the novel technologies and requirements of the modern economy, which warrant quality trainings for the instructors.
- Skills training programmes in India often neglect the so-called "soft skills" and the lack of such skills can hamper career advancement.
- Vocational training in India is associated with low-prestige jobs and is frequently not considered as a viable alternative to formal education. Vocational education has to be made more attractive with better links to competitive jobs with higher salaries.
- A labour market information system is critical for aligning demand and supply for occupations and skills, which is not fully achieved.
- Financing institutions are not willing to provide loans or microfinance instruments if employment is not guaranteed
- Indian companies cite a number of challenges that hamper the participation of working-age women in skills development opportunities and in the labour force.

In the context of India's booming gig and platform economy, there are challenges that need to be also addressed. Platforms have given rise to innovative models that can provide social security benefits to workers independent of their association to platforms. The best practices from countries such as the US, the UK, Singapore, China, South Africa and Hong Kong have been used as evidence to underscore the need for creating specific policies to promote platform economy and protect the platform workers. India's Code on Social Security expands worker classifications in India, recognises worker classifications of gig and platform workers and extends social protection to all workers. Extending social security to those associated in non- traditional work relations such as platforms is key to universalising social security

3.8. Addressing the key challenges

As structural, demographic and technological shifts transform the Indian economy and the nature of work, new entrants to the labour force will have to be skilled and made employable. Strategies for reskilling and enhancing the skill set of the current workforce, as well as formal recognition of informally acquired skills, will have to be reinforced.

Appropriate actions and strategies have to be adopted to address the three key challenges to skills development in India: expanding public sector collaboration with industry and the private sector, creating pathways for international mobility and addressing women's low participation in the labour force.

1. Industry-Private Sector Collaboration

Creating possibilities for private sector engagement has been one of the crucial strategic pillars for skill development in India. The skilling landscape faces an acute gap in terms of sectoral need and availability, competency required by an employer and those possessed by a trainee. Basically, there is an issue of information asymmetry in multiple fronts. Recognition of Prior Learning (RPL) is one promising way to counter the same.

Additionally, with Industry 4.0 gaining traction over years, training programmes to up-skill and reskill the workforce of the future need to be more scalable, personalized, and flexible to adjust to the dynamic needs of the massive digital revolution. However, at this point, not many companies may be able to predict the exact industry-specific and occupation-specific competencies that they need few years down the road. The issue is further aggravated by the lagging pace of industry growth due to traditional education curricula and testing methods, which has been making it imperative for the industry to collaborate with professional skill training institutions.

In the recent years, professional learning programmes and apprenticeships have become one of the highly popular ways among global companies to address the skill gap, as well as curb unemployment in the country. In India, the Automotive Skill Development Council (ASDC) has been a key establishment for the India Auto Inc. that has been supporting the industry by continuously developing data-driven solutions to the existing skills challenges. Apart from offering massive training programmes aimed to reskill and up-skill the auto workforce, ASDC provides an inventory of skilled manpower in the automotive sector, which includes candidates who have undergone skill training and assessments. Agricultural Skill Council of India (ASCI) is also partnering with corporates for boosting the digital revolution in the agricultural sector. However, there is more that needs to be addressed.

Skill development faces several forms of market failures, including information asymmetries – a skilled person knows his or her skills, but a potential employer does not; if employers had all the information, their willingness to pay for a skilled person would rise. Recognition of Prior Learning (RPL) is an example of an intervention to address information asymmetry. Another market failure in skill development is externalities. For example, when a firm spends resources on developing the skills of an employee, he or she can quit and join another firm, thus benefitting the latter firm and not the firm that incurred the training cost.

In India, there are a variety of skill development models – government-funded programmes that fully or partially subsidize training/apprenticeships, market-led trainings (where trainees pay for the course), market-led apprenticeships and industry-led/on-the-job training. The not-for-profit National Skill Development Corporation (NSDC) was set up as a public-private partnership (PPP) to stimulate private sector participation in the Indian skill development sector.

Sector Skill Councils (SSCs) were incubated by the NSDC for fostering industry connections and developing an industry-relevant course and curriculum. There are currently 37 SSCs operational, with more than 600 corporate representatives in their governing councils.

2. Encouraging international mobility

India is firm at its stance on becoming the 'Skill Capital' of the world, and the commitment is made evident through its structured efforts, including the India International Skill Centre (IISC) programme. For the purpose of counselling and guidance to the potential emigrants, with special focus on skills tests, upskilling, language and pre-departure orientation, a proposal for a new and market-driven IISC network has also been initiated. Under Government-to-Government (G2G) and business to business (B2B) partnerships, India and Japan are cooperating to implement Japan's Technical Intern Training Programme (TITP), an on-the-job training scheme providing three to five years of internship opportunities for foreign nationals in Japan. Furthermore, the government has undertaken technical collaborations with countries including the UK, Australia, and the UAE for benchmarking, and mutual recognition of standards.

In the Indian Auto sector, ASDC has joined hands with Magic Billion, a leading International Skill Centre under Skill India Programme, in an attempt to further the PM's vision of "Making India the Skill Capital of the World". With all the new-age vehicles technologically more advanced, connected and smart, it calls for an entirely new set of skills and knowledge to drive the industry growth and its complete value-chain. The projects undertaken through this collaboration also demonstrates that ASDC is creating career pathways for professionals in Automotive Sector, by initially training them on Indian Qualifications and then sending the best to global careers.

Technical collaborations have been undertaken with countries such as the UK, Australia and the UAE for benchmarking and mutual recognition of standards. G2G and B2B partnerships are also being developed for new markets such as those in Western Europe, Canada, Australia and East Asia to increase the mobility of blue and white-collar Indian workers.

3. Low participation rates of working age women in the labour force

Another big challenge is the low rate of female participation in the labour force. According to various reports, out of the country's labour force of 395.2 million, only 91.6 million are women. Apart from various skilling initiatives undertaken by the government to curb the gap, complemented by a wider push towards empowerment through gender sensitization, creation of economic opportunities and economic and social support, the focus should be on creating a gig economy, and provide flexible work models to attract more women. Given that a majority of women respondents report 'attending domestic duties' as their reason, flexibility in the system might be helpful.

Fortunately, the country's progress so far, in terms of creating access to skill development for women has been fruitful, with currently, more than 50 per cent of the total candidates trained under PMKVY, being women. There is even a growing trend of women increasingly getting trained and enrolled for unconventional job roles in automotive, electronics, and hardware sectors. The government is also encouraging partnership explorations with industries to support women-centric projects in non-traditional trades.

Skilling initiatives complemented by a wider push towards empowerment through gender sensitization, creation of economic opportunities and economic and social support can be effectively utilized to raise participation of women. Providing residential facilities for women trainees, embedding mentoring and coaching in skills programmes and providing social support through mechanisms such as local workshops need to be explored further.

Besides the above, there is an increasing size of gig workforce in India and there need to be policy formulations to unlock jobs in this area with emphasis on identifying and removing entry barriers for totally democratising access to jobs in this domain. The role of digital platforms in skill development needs special focus and attention. The policy also needs to be formulated to upskill the gig and platform workforce and provide them social security measures.

3.9. Improving Employability

Improving employability through higher education remains a challenge and can be attributed to many factors. A review of the market trends and research broadly classifies the issue into three broad factors, as below:

- Skill mismatch
- Quality skill gap
- Over-qualification

It is quite unfortunate that the next generation of the workforce in India continues to be trained using a curriculum that has remained vastly unchanged for several years. The post-COVID era further warrants realignment of the higher education curriculum, up-skilling practices, and training techniques. Dynamic changes to the jobs market as a whole and sector-specific requirements further results in recruiters demanding updated and evolving higher order curriculum and candidates with multidisciplinary skillsets.

The recruiters and organizations are keen to identifying more adaptable, solution-oriented and application-focused freshers. However, inclusion of these professional skills has not ideally materialised in the higher education curriculum nor has its integration commenced with the technical training received by the higher education learners.

Technical skills often acquired by learners in various Higher Education (HE) programmes are often not required for several jobs in the market. In addition, employers continue to stress that technical skills, professional skills, or knowledge are often inadequate for the job role requiring HE graduates with specific training.

The skill gap is further exacerbated by the post-COVID jobs market trends. The recruiters are increasingly looking for multi-talented hires, who are skilled in more than one technical competency and can supplement with strong professional skills, such as critical thinking, problem solving, adaptability, leadership, and communication skills. In addition, the scarcity of graduates specializing in programmes or technical skills that are in high demand continues to exacerbate the industry and higher education skill gap. For instance, the demand for competencies in data analysis, strategy, technology, and problem solving continues to grow.

References

AFC India 2025 Skill and Human Capital development, https://www.afcindia.org.in>skill_and_human_capital_development

Mehrotra, S and Saxena, P K 2014 The Skill development system in India: A Reform Agenda, https://doi.org/10.1093/acprof:oso/9780199452774.003.008

MOSPI 2022 Annual Report PLFS 2021-22, https://www.mospi.vov.in>publication>annual-report

Nick O'Donovan 2022 Pursuing the Knowledge Economy: A Sympathetic History of High Skill, High wage hubris, Agenda Publishing, New Castle upon Tyne

NITI Aayog 2022 India's Booming Gig and Platform Economy: Perspectives and Recommendations on the Future of Work, Report, June 2022

Osborn, J 2020 Platform-led upskilling in Africa: Can Digital Platforms help close the Continent's Skill gap? NextBillion, March 10, 2020.

Sharma, Dewangi 2024 Skill India: Education and Employment Bridge, Jan 4, 2024, https://www.investindia.gov.in>blogs>skill-india-education

Urbane Parent 2024 Career and Vocational Course in India, https://urbaneparent.com>blog>vocation-and-career

World Economic Forum, 2019 The three challenges to Skill development in India and how to tackle them, https://www.weforum.org>stories>2019/10>india-skill

CHAPTER 4

POLICY INITIATIVES FOR SKILL DEVELOPMENT IN INDIA

India is one of the youngest nations in the world with more than 54 percent of the total population below 25 years of age, the India's workforce being the second largest in the world after China. While China's demography dividend has started tapering off presently, India will continue to enjoy it till 2040 (GoI, 2016). However, India's formally skilled workforce is only approximately 2 per cent, which is dismally low compared to other countries. To influence our demography dividend more substantially and meaningfully, the government launched the 'Skill India' campaign along with 'Make in India'.

Skill development plays a vital role to reduce poverty by improving employability, productivity and helping sustainable enterprise development and inclusive growth. It facilitates a cycle of high production, increased employment opportunities, income growth and development. Skills and knowledge are the driving forces of economic development of any country. They have become even more important given the increasing pace of globalization of world economy and technological changes.

India is currently the fifth largest economy in the world according to the IMF data based on growth and nominal GDP and is on track to become third-largest by 2030 and second-largest by 2075. The median age in India is 29 years against the average age of 37 years in China and the US and 45 years in Western Europe implying that India has a huge demographic dividend. A well-designed and implemented skill development policy is essential for India to reap the benefits of its demographic dividend. India's young population must be equipped with the skills necessary to compete in the dynamic global economy.

With the present education and skill levels of those already in the labour force being at a very low level, it would be a major challenge for India to reap its demographic advantage. To add to this, the challenges of an open economy call for constant re-skilling and up-skilling of the nearly 40 per cent populations in the 30-59 age group to maintain their employability levels.

Since independence, India has come a long way in upskilling its population. Industrial policies have played an important role in successful economic development in various countries. Through these policies, governments intervene in the market's sectoral allocation of resources and choice of technologies.

4.1. Industrial Policies

The Industrial policy of 1948 was first major policy of independent India launched to lay the foundation of a mixed economy in which both private and public enterprises would march hand in hand to accelerate the pace of industrial development.

Industrial Policy Resolution (IPR 1956) adopted by the Indian Parliament in April 1956 was the second comprehensive statement on industrial development of India after the Industrial Policy of 1948. The 1956 policy continued to constitute the basic economic policy for a long time. According to this resolution, the objective of the social and economic policy in India was the establishment of a socialistic pattern of society, that provided more powers to the government machinery. It laid down three categories of industries which were more sharply defined. These categories were:

- Schedule A: those industries which were to be an exclusive responsibility of the state.
- Schedule B: those which were to be progressively state-owned and in which the state would generally set up new enterprises, but in which private enterprise would be expected only to supplement the effort of the state; and
- Schedule C: all the remaining industries and their future development left to the initiative and enterprise of the private sector.

Realising that no country has achieved sustainable economic and human development without significant and continuous investment in the knowledge and skills of its people, the industrial policy has given a bigger role for skill development.

4.2. Apprenticeship Act

Apprenticeship is a skill training programme wherein a person is engaged by a company or firm as an apprentice and gains classroom (theory) learning for a short period, followed by on-the-job (practical) training. An apprentice is treated as one who has signed a contract of apprenticeship with an establishment to undergo apprenticeship training.

Acknowledging that training imparted in institutions are not sufficient for acquisition of employable skills and needs to be supplemented by training at the workplace, the Apprentices Act, 1961 and Apprenticeship Rules,1962 were enacted with the prime objective to utilize fully the facilities available in industries for imparting practical training and thus developing skilled manpower for industries. Initially, the Act covered the apprenticeship training for the trade Apprentices. Subsequently, the Act was amended in 1973, 1986 and 2014 to bring the Graduates, Technician, Technician (Vocational) and Optional Trade Apprentices respectively under its purview and accordingly, Apprenticeship Rules, 1992 were revised in 2015.

4.3. National Apprenticeship Promotion Scheme (NAPS)

NAPS was launched in August 2016 by Government of India to promote the apprenticeship in the country by providing financial incentives, technology and advocacy support. The scheme has two components:

- Sharing of 25 per cent of prescribed stipend subject to a maximum of Rs. 1500/- per month per apprentice with the employers.
- Sharing of basic training cost up to a maximum of Rs. 7,500 per apprentice.

The key features of NAPS are:

1. Wider option for the apprentices integrating with other schemes : The courses approved by State Government/Central Government such as PMKVY, DDU-GKY etc. can be linked with apprenticeship training. These courses will be given status of optional trades and the relevant practical content for on-the-job training will be added by respective course approving authority.

2. Ease of Administering through technology: A specially designed online portal www.apprenticeshipindia.org is used for administering the entire implementation of the Apprenticeship Training online. It facilitates the requirements of all key stakeholders such as Candidates, Industry, DGT, RDSDEs, NSDC, SAA, SSDMs and BTPs.

3. Involvement of States/UTs: As per the Apprentice Act, monitoring of apprenticeship training in State Public Sector Undertakings and Private Sector Establishments is done by the respective State Governments. The states have a very important role in implementing the programme as most of the smaller industries and MSMEs come under the State jurisdiction. Hence sensitizing the states and State Skill Development Missions (SSDMs) and also officers up to the district level, on the intent of the reforms becomes critical and a matter of priority. The District Skilling Committees promoted by both the Central and State governments are mandated to identify apprenticeship opportunities in their district and ensure appropriate utilisation of same.

4. Promoter and Facilitators/Third Party Aggregators (TPA): Since this scheme involves multiple stakeholders, the role of facilitators or Third Party Aggregators (TPAs) becomes important for mobilizing the apprentices, mapping their preferences with the demand from the establishments for apprenticeship opportunities posted on the portal, and helping the establishment in identifying Basic Training Providers. TPAs are engaged as per the guidelines issued by MSDE for their selection.

Implementing Agencies for NAPS

The Regional Directorates of Skill Development and Entrepreneurship (RDESE) under Directorate General of Training (DGT) are the implementing agencies in their regions in respect of all 'Designated Trades' under the Act for all establishments coming under the Central Government jurisdiction. National Skill Development Corporation (NSDC) and CEOs of the Sector Skill Councils are the implementing agencies in their sectors in respect of 'Optional Trades' for the establishment under the Central Government jurisdiction.

The respective State Governments are the appropriate authorities in respect of any establishments other than those falling under the Central Govt. jurisdiction. The State Apprenticeship Advisers (SAA) are implementing agencies in their regions in respect of all 'designated trades' a well as 'optional trades' for State Public Sector Units and private establishment falling under their jurisdiction as per the Apprentices Act 1961. They may also appoint the 'Mission Directors' of the respective State Skill Development Missions (SSDMs) to act as implementing agency for all establishments under the State Government jurisdiction, in respect of 'Optional Trades'.

4.4. Indian Education Commissions

In the independent India, vocational education received policy attention in the University Education Commission in 1948 (Chairman: Dr. S. Radhakrishnan) and Secondary Education Commission in 1952 (Chairman: Dr. L. Mudaliar). The first one (1948) recommended that in order to direct students to vocational training at the end of class/grade 10, a large number of intermediate colleges should be opened. These colleges were to meet variety of needs of young men and women by giving a vocational bias to their courses while retaining their value in a system of general education as preparation for university courses. The Secondary Education Commission (1952) recommended that at the end of secondary education the student should be in a position by choice to enter into world of work taking up some vocations. It resulted in creation of multipurpose schools. The Education Commission of 1964 (Chairman: Dr. D.S. Kothari) suggested restructuring of Education to 10+2+3 pattern and district streams of general and vocational education at the higher secondary stage were recommended with an intention of intercepting the goal-less climb-up of the youth on the educational ladder and diverting them to a productive path.

The Kothari Commission (1964-66) believed that *the Indian education system did not prepare learners to take a career path.* The recommendation reads, 'In order to increase national income in a country like India, education should be related to productivity. The quality of science teaching should also be raised considerably to achieve its proper objective such as develop problem solving skills, analytical skills and their easy application to the problems of the material environment and social living. Another programme which can bring education into closer relationship with productivity is vocationalization.'

The Commission recommended development of a comprehensive system of vocational education in relation to the developmental needs of the country. The Commission suggested the establishment of vocational training institutions to provide skill-based training to the youth. It suggested that such institutions should be established in rural areas to provide technical and vocational education to the rural population.

4.5. National Policies on Education

The recommendations of Education Commissions became a vital component of the first National Policy on Education (NPE 1968) where in the "effectively terminal nature of the vocational stream" was emphasized. In 1968 the Government of India formulated the National Policy on Education, in response to the recommendations of the Kothari Commission. The policy sought 'total reformation' and aimed at extending the prospects of education to all sections of the society to accomplish the goal of harmony and integration. The policy suggested the provision of compulsory education to children in the 6-14 years age group as proposed in the Indian Constitution.

The second National Policy on Education (1986) has accorded high priority to the programme of Vocationalisation of Education. The introduction of systematic, well planned and rigorously implemented programmes of vocational education was highlighted in the proposed educational reorganization. It stated that the vocational education will be a district stream, intended to prepare students for identified occupations spanning several areas of activity and that vocational courses will ordinarily be provided after the secondary stage, but keeping the scheme flexible, available after class VIII.

The revised policy formulations (1992) retained policy framework laid by NPE 1986. Moreover, it envisages, children at the Higher Secondary level are imparted generic vocational courses which cut across several occupational fields and which are not occupation- specific.

The National Education Policy (2020) lays particular emphasis on the development of the creative potential of each individual. It is based on the principle that education must develop not only cognitive capacities, both the 'foundational capacities 'of literacy and numeracy and 'higher-order' cognitive capacities, such as critical thinking and problem solving, but also social, ethical, and emotional capacities and dispositions.

The pitfalls of NEP 2020 include increased student pressure, narrowing of subject choices, insufficient teacher training, digital infrastructure imbalance, standardisation challenges, and an insufficient implementation strategy.

The National Education Policy (2023) introduces reforms such as an extended schooling period (ages 3-18), a shift in curricular structure, emphasis on native languages, higher education reforms, blurring of subject distinctions, flexible undergraduate degrees, universal access to school education, foundational literacy and numeracy, experiential learning, increased graduate enrolment, open and distance learning enhancement, technological innovation, internationalisation of education, and increased public investment.

The new education policy addresses issues by connecting vocational and formal education, introducing flexible structures, promoting experiential learning, enhancing technological innovation, encouraging international collaborations, and increasing public investment to improve educational infrastructure.

4.6. National Commissions on Labour

The First National Commission on Labour (1969) recommended that there should be a policy framework in the unorganized sector that ensures the generation and protection of jobs, access to jobs, protection against the exploitation of poverty and lack of organization, protection against arbitrary or whimsical dismissal and denial of minimum wages. It also suggested that a system of welfare should include access to compensation for injuries suffered while engaged in work, provident fund, medical care, pension benefit, maternity benefits, child care shelter and old age support.

The Second National Commission (2000) recommended that the Training and Employment Organisations in the State should be under the State Labour Department and there should be uniform minimum qualifications and comparable enhancement prospects for instructional staff and uniform vocational standards of training in all States. A sustained publicity of the programmes and achievements of the employment and training services should be organised in order to convey their full importance to employers/ trainees and the public.

The main task of training of workers should necessarily be vested in the industry. The States should supplement rather than supplant the activities of employers in this matter. It should step in only in such fields and areas where employers cannot undertake training programmes. The State Apprenticeship Adviser should be appointed as the Authority for registering the training schemes organised by employers.

4.7. National Policy on Skill Development

The Taskforce on Skill Development set up by the Planning Commission in 2007 emphasised that 'the present state of affairs is not acceptable for a country which seeks to become a knowledge power, a global service and manpower provider and a leading industrial nation. There has to be a paradigm shift in the national policy on skill development with the private sector playing a lead role instead of the government, as they are the job providers. The government's role would have to change from being a vocational training provider to a partner and facilitator and it is this philosophy which underpins the Task Force Report'.

The Report of Task Force on Skill Development (PC, 2007) noted the creeping urbanization of India gathering momentum and is irreversible. The indication was clear that unskilled manpower is available and the skill shortage is acute. This paradox has to be reconciled, and vocational training should be made available to those who seek it. As training cannot thrive in isolation, it is imperative that the employers and NGOs come aboard, as equal partners, and help run the new institutions like the PPP- based ITIs and the Skill Development Centres. The trainees would not just come from the organized sector, but also the unorganized or informal sector, in semi-urban and rural areas.

The Eleventh five-year plan (2007-2012) marked the beginning of a structured approach to a skill development programme. This resulted in the creation of the Co-ordinated Action on Skill Development in 2008 followed by the formulation of the National Policy on Skill Development (NPSD) in 2009. A three-tier governance structure primarily led by the government and other public-sector bodies was also rolled out.

The National Policy on Skill Development formulated in 2009 provided the framework for skill development activities in the country. However, the changes in the macro environment, and the experience gained through implementation of various skill development programmes in the country over the years have necessitated changes in the policy. Accordingly, the National Skill Development Policy, 2015 was formulated, and it supersede the Policy of 2009.

The shift in policy vision from undertaking skill development primarily through vocational training to encouraging innovations and entrepreneurship in an outcome-oriented framework is clearly observed comparing the National Policy on Skill Development (NPSD) 2009 and the NPSD (Entrepreneurship) 2015. However, the blueprint of the organization of the skill development programme has been carried forward as conceived first in 2009 under the NPSDE. Much of the programme elements crafted by the policymakers in the UPA government and those which could not be realized during their term have been adopted by the NDA government.

Comparison of NPSD, 2009 and NPSDE, 2015

	NPSD, 2009	**NPSDE, 2015**
Vision	Initiative -Empowering individuals through access to better skills & knowledge conforming to qualification standards - Enhancing access to decent employment and India's competitiveness in the global market	Mission - Focus on sustainability of livelihoods by accelerating skill development on a large scale with focus on standards - Promote and nurture innovations and entrepreneurship unfolding newer avenues of employment and wealth creation
Mission	Rapid & Inclusive Growth - Enhance employability - Adaptation to changing technologies and labour market demands -Improve productivity and living standards - Strengthen India's competitiveness - Attract investment in skill development	Dynamic entrepreneurial economy - Create demand for skilling - Correct and align skilling with required competencies - Connect supply of skilled human resources with sectoral demands - Align national and global certification and assessment standards - Catalyze sustainable ecosystem for innovation and entrepreneurship
Objective	Create opportunity for Skilling for all individuals - Emphasis on skilling of youth, women & disadvantaged groups, promote stakeholder commitment, align supply of skilled workforce with sectoral demand	Make skilling aspirational for youth and employer - Seamless integration of skilling with formal education - Improve capacity and quality of infrastructure and resources - Align supply with Make in India initiative, in addition to sectoral needs

The NPSDE 2015 aimed at a more integrated and holistic approach for attaining skill development with greater emphasis on demand-driven outcomes. It also suggested to streamline the role of different ministries and public agencies at the central and state levels to avoid duplication of efforts and to achieve diversification of human and financial resources. The skill development programme is linked with other flagship programmes such as Digital India and Make in India since skilled manpower is the key driver for these programmes. The NPSDE 2015 also lays down diversified financing sources for skill development programmes in addition to the public spending undertaken by the central government.

The **vision** of NPSDE 2015 is to create an ecosystem of empowerment by Skilling on a large Scale at Speed with high Standards and to promote a culture of innovation-based entrepreneurship which can generate wealth and employment so as to ensure Sustainable livelihoods for all citizens in the country.

The **core objective** of the Policy is to empower the individual, by enabling her/him to realize their full potential through a process of lifelong learning where competencies are accumulated via instruments such as credible certifications, credit accumulation and transfer, etc. As individuals grow, the society and nation also benefit from their productivity and growth.

4.8.Policy framework for Skill Development and Entrepreneurship

The policy framework has been developed to accomplish the vision of Skill India by adhering to the core objective. The framework outlines eleven major paradigms and enablers to achieve the objectives of skilling India. They are:

- **Aspiration and Advocacy:** The need of the hour is to make skill development aspirational for boys and girls in our country. State-wise camps will be organised for skill awareness and mobilization.
- **Capacity:** The government will support the creation and use of infrastructure in both public and private domains through appropriate equity, grant and loan support.
- **Quality:** Quality of training can be measured by competency outcomes and employability of trainees. The national standards and quality for skilling are globally aligned and Indian youth can aspire to secure local, national and international job opportunities.
- **Synergy:** MSDE has been set up to co-ordinate skill development efforts in the country. The national Labour Market Information System (LMIS) created is an integrated database.
- **Mobilization and Engagement:** Industry will be encouraged to increasingly move towards employing only certified skilled people. Industry should actively participate in designing curricula and standards for skill training

courses. Since industry is one of the recognized stakeholders, it needs to actively contribute to the cause of skill development.

- **Global Partnerships:** The main objective of global partnerships and international collaboration is to leverage best practices from across the world. Foreign governments, corporates and agencies will be encouraged to set up skill centres and universities as well as participate in content creation, design of curricula and delivery of training.
- **Outreach:** More than 93 per cent of India's workforce is in the unorganised sector. Hence strengthening and certifying the skills of the unorganised workforce will contribute to overall economic development of this sector.
- **ICT Enablement:** The use of technology will be leveraged to scale up training facilities, enable access to remote areas and increase cost effectiveness of delivery of vocational trainings. Stakeholders will be encouraged to develop MOOC and virtual class rooms for easy access and convenience.
- **Trainers and Assessors:** To achieve the massive target of skilling, it is important to have quality trainers who are capable of training people in several domains. Similarly, quality assessors in sufficient numbers are also required to ensure consistent outcomes of assessment and certification process.
- **Inclusivity:** One of the key objectives is to safeguard the skilling needs of SC, ST, OBC, minorities and differently abled persons as well as those living in difficult geographical pockets.
- **Promotion of skilling among women:** Women participation in vocational education and training is especially low as compared to men. The Government will promote setting up of more training and apprenticeship seats exclusively for women. Appropriate incentive mechanism will be designed to achieve the same.

Recognizing the need to better coordinate the diverse efforts of various ministries and government departments in promoting skill development, the Government notified the formation of a Department of Skill Development and Entrepreneurship in July 2014 under the Ministry of Youth Affairs and Sports, which was later upgraded to Ministry of Skill Development and Entrepreneurship (MSDE) in November 2014. MSDE is an umbrella Ministry coordinating all activities around skill development for not just achieving the ambitious targets set out for skilling people but for creating a national ecosystem of skilled workforce and fostering innovations and entrepreneurship powered by state-of-the-art infrastructure, high quality standards and mechanisms for compliance.

With the announcement of the NPSDE in July 2015, the National Skill Development Mission (NSDM) was approved and the different committees under the Ministry were notified.

4.9. PMKVY

Among the various skill development programmes spread across 20 other departments and Ministries, Pradhan Mantri Kaushal Vikas Yojana (PMKVY) is a flagship programme of the MSDE, GoI launched in July 2015. PMKVY was launched to encourage and promote skill development in the country by providing free short duration skill trainings and incentivizing this by providing monetary rewards to youth for skill certification.

After its successful implementation in 2015, the timeframe of the programme was extended up to 2022 with the ambitious target of skilling 10 million youth. Being a government funded programme, the PMKVY is being implemented by the NSDC under the guidance of the MSDE.

After the successful implementation of pilot PMKVY (2015-16), PMKVY 2.0 was launched by scaling up both in terms of Sector and Geography and by greater alignment with other missions of Government of India like Make in India, Digital India, Swachh Bharat, etc.

The scheme is being implemented through two components:

1. **Centrally Sponsored Centrally Managed (CSCM):** This component is implemented by National Skill Development Corporation. 75 per cent of the PMKVY 2016-20 funds and corresponding physical targets have been allocated under CSCM.
2. **Centrally Sponsored State Managed (CSSM):** This component is implemented by State Governments through State Skill Development Missions (SSDMs). 25 per cent of the PMKVY 2016-20 funds and corresponding physical targets have been allocated under CSSM.

Based on the learning from the implementation of the schemes, the Ministry had launched the third phase of Pradhan Mantri Kaushal Vikas Yojana (PMKVY 3.0) in January, 2021 to train 8.00 lakh youth across the country. Further, the Ministry had launched two special programmes under PMKVY 3.0, viz. Customised Crash Course Programme for COVID Warriors (CCCP for CW) to mitigate the impact of COVID-19 pandemic and Skill Hub Initiative (SHI) for integration and mainstreaming of vocational education with general education as envisaged under the National Education Policy, 2020.

PMKVY 4.0 is launched from February, 2023 designed to make the scheme flexible, inclusive, technology enabled, and market driven resulting in better employability. The broad design principles of PMKVY 4.0 are:

- Enhance industry-connect, greater reliance on On-the-Job Training in short-term training for learning by doing. In addition, encourage industry buy-in and contribution for skilling including sharing of training infrastructure, demand aggregation, OJT, etc.
- Emphasis on re-skilling, and upskilling under Recognition of Prior Learning (RPL) through innovation in assessment and better monitoring.

- Digitalization of training ecosystem through online integrated systems including registration and counselling,
- Delinking of placement from skilling to lay emphasis on empowering trained candidates to choose their career path with focus of entrepreneurship and livelihood creation.
- Flexibility in course curriculum by introducing courses in partnership with industry, Ministries/Departments, etc.
- Focus on online/digital/blended skilling for effective learning, convenience, and outreach.
- Focus on New Age Skills like Industry 4.0, Web 3.0, AI/ML, AR/VR, Climate Change, Circular Economy, Green Economy, and Energy Transition

The implementation of PMKVY 4.0 scheme typically involved the following steps:

- Design and Planning and identify the target sectors
- Training Provider Selection
- Curriculum Development
- Training Delivery
- Assessments and Certifications
- Recognition of Prior Learning (RPL)
- Placement Assistance
- Monitoring and Evaluation of the training programme

4.10.Skills for Migrant workers

The relationship between migration and skill development proves to be a complex one. Migration can be a means to respond timely and effectively to labour supply and demand needs. Migrants, at all skill levels, broaden the pool of available skills, thereby improving the internal flow of talent and economy wide skills matching.

The COVID-19 pandemic brought with it several challenges for India and nations across the world. One of the worst affected was the livelihoods of many unorganized and migrant workers. In this context, several initiatives were undertaken by states to manage the movements of migrant workers across districts, states and regions. The workers had to be provided with food, shelter and alternative sources of livelihood. It is worth appreciating that several states took swift measures in assisting stranded migrant workers and launching initiatives in the areas of skill-mapping and livelihood generation for those who had returned to their states of origin and other workers in the unorganized sector.

NITI Aayog (2020) has brought out a publication titled 'Empowering migrant workers through skill development and livelihood generation: Best practices from States and Districts'. This compilation of best practices from across states is a glimpse into the massive and relentless exercise undertaken by various state governments to address grave challenges in the wake of the COVID-19 pandemic, particularly through skilling.

The Ministry of External Affairs (MEA) set up the India Centre for Migration (ICM) as a 'not for profit' society in July 2008. The centre is the implementing partner representing Ministry of External Affairs (MEA) in the technical and support project under the aegis of the Joint Declaration on India-EU Common Agenda on Migration and Mobility (CAMM). ICM plays a pivotal role in conducting Training of Trainers (ToT) workshops and certifying Master Trainers. The trainers impart Pre-Departure Orientation (PDO) training to migrant workers in order to ensure that workers migrate safely for overseas employment with an enhanced skill set.

4.1 New Labour Codes

With the introduction of the new labour codes, India aims to strike a balance between enhancing labour market flexibility and improving workers' rights and welfare. This balance will have significant implications for both domestic and foreign businesses.

The four labour codes—Code on Wages, Industrial Relations Code, Social Security Code, and Occupational Safety, Health and Working Conditions Code—aim to simplify existing laws, making compliance easier and protecting workers' rights under a cohesive regulatory framework. The new codes also decriminalize minor offenses, emphasize skill development, and streamline dispute resolution mechanisms.

The Code on Social Security (2020): Focuses on expanding social security benefits to all workers, including gig and platform workers. It introduces a Re-skilling Fund and allows the government to extend benefits through the Employees' State Insurance Corporation (ESIC) and Employees' Provident Fund Organization (EPFO).

Source: https://labour.gov.in>labour-codes

Policy Framework for Entrepreneurship

The entrepreneurship policy framework has been developed considering the fact that vibrant entrepreneurship requires support from an enabling ecosystem of culture, finance, expertise, infrastructure, skills and business friendly regulation. Many government and nongovernment organizations are playing enabling roles across each of these crucial supporting elements. This policy framework, cognizant of the need for the full ecosystem to be present to unlock entrepreneurial potential, proposes a nine part entrepreneurship strategy:

- Educate and equip potential and early stage entrepreneurs across India

- Connect entrepreneurs to peers, mentors and incubators.
- Support entrepreneurs through Entrepreneurship Hubs (E Hubs).
- Catalyse a culture shift to encourage entrepreneurship.
- Encourage entrepreneurship among underrepresented groups.
- Promote entrepreneurship amongst women
- Improve ease of doing business.
- Improve access to finance.
- Foster social entrepreneurship and grassroots innovations

4.11. Skills and Agri Entrepreneurship

The Entrepreneurship and Skill Development Programme (ESDP) is a Central Sector continuing scheme, under the Ministry of Micro, Small and Medium Enterprises(MSME). Aspiring and existing entrepreneurs, MSME owners aspiring to scale their businesses, individuals from weaker sections interested in starting a business, etc can join the scheme. The qualification of the participants and structure of the fees will be decided by the Implementing Agencies. The age of the participants will be 18 years and above.

ESDP

The objective of the programme is to motivate youth from different sections of society including SC/ST/women, differently-abled, ex-servicemen and BPL persons to consider self-employment or entrepreneurship as a career option. The ultimate objective is to promote new enterprises, ensure capacity building of existing MSMEs and float an entrepreneurial culture in the country. The activities have been undertaken in 17 states, namely: Assam, Tamil Nadu, Madhya Pradesh, Kerala, Karnataka, Chhattisgarh, Gujarat, Haryana, Jammu and Kashmir, Jharkhand, Maharashtra, Odisha, Punjab, Telangana, Uttar Pradesh, West Bengal, and Uttarakhand. The initiative comprises the following modules:

1. Industrial Motivational Campaign (IMC)
2. Entrepreneurship Awareness Programme (EAP)
3. Entrepreneurship-cum-Skill Development Programme (E-SDP)
4. Management Development Programme (MDP)

IMC leads to increased awareness among the participants about prospects of entrepreneurship, procedures of creating new enterprises and different government schemes, funding opportunities, etc. The youth and prospective entrepreneurs are motivated to set up micro-enterprises in traditional/non-traditional domains.

EAP trains the participants about entrepreneurship in general and procedures of starting an enterprise in particular. They are also exposed to different government schemes, funding opportunities, entrepreneurial risks and returns.

The participants develop the motivation to start new enterprises. The participants attending the programme get a clear and executable business plan enabling them to undertake measures for launching their enterprises.

E-SDP enhance awareness and knowledge of participants about various entrepreneurial aspects and skill domains and policies and schemes of the Government, funding opportunities, calculated risks, Business opportunities, Business Plan Preparation and related areas. The programme also aims at imbibing learning on new skills and trades. Thus, this programme ensures that the participants use their entrepreneurial as well as technical skills to set up sustainable enterprises.

MDPs aim at improving managerial and organisational efficiency of entrepreneurs and managers for running their enterprises.

These programmes are conducted with the support of Ministry of MSME, Govt. of India.

Government Schemes for Entrepreneurs

1. Small Scale Industries

The growing SME sector of India propels a portion of the development of the economy. The small and medium-sized sector accounts for over 40 per cent of the total GDP and continues to be a vital source of jobs for the increasing population of India. Recognising the significance of local SMEs, the government has launched several schemes to boost the SME sector.

As a result, India is experiencing a boom in startups in all industries, including IT, service delivery, wellness, technology, and others. The government has launched several schemes to make the process fast in order to enable young entrepreneurs to kickstart startup companies that will ultimately generate job opportunities. The majority of these schemes offer financial aid in the form of subsidies and loans to prospective individuals and organisations.

2. Budding Entrepreneurs

i. Atal Incubation Centre (AIC)- It is a funding scheme started in 2016. Envisioned by the NITI Aayog, this aims at supporting entrepreneurs by covering their operating costs of capital. The approved business ventures can avail up to Rs. 10 crore over a five-year term. AIC allows researchers, students, and startup owners to apply for the scheme across different verticals and sectors. The entrepreneurs can set their company as a Public-private partnership, Public organisations, or as a completely private organisation.

ii. The MSME Business Loans in 59 Minutes- It is undoubtedly the most promising business loan scheme introduced in 2018 by the Government of India to provide financial assistance to micro, small, and medium enterprises. The entrepreneurs can avail of loans up to one crore under this scheme. This scheme is very quick and, within 59 minutes, one can know about the eligibility. The disbursement of the financial assistance will take place within one to two weeks. A majority of the public sector banks are a part of this scheme.

iii. National Small Industries Corporation Subsidy- NSIC subsidy offers two forms of financial benefits for small businesses: Raw Material Assistance and Marketing Assistance. Raw Material Assistance helps to procure raw materials from abroad and from India. Marketing Assistance, on the other hand, helps to improve the sales of products and services with an efficient marketing process.

iv. MSME Market Development Programme- This is designed to provide market expansion assistance to micro, small, and medium enterprises using international events, trade fairs, and roadshows. This system helps in the growth of the business by providing them assistance in terms of expansion into international markets. Any startup organisation registered under District Industries Centre can apply for this programme.

v. MUDRA Loans- The Government of India initiated MUDRA loans for providing business finance for micro-business units. This programme was launched with the single motive of 'paying the unfunded'. Since most of the time, the small-medium enterprises suffer from lack of funds, MUDRA was launched to encourage participation and growth of the startups across different sectors like trading, manufacturing, services, etc.

vi. The Women Entrepreneurship Platform- The Government of India introduced the Woman Entrepreneurship Platform to promote women's entrepreneurship. The NITI Aayog is the primary driver behind this programme, which aims to boost the morale of young and dynamic woman entrepreneurs. This scheme has got three divisions: Gyaan, Iccha, and Karma Shakti. Iccha Shakti aims to empower women to start a company. Gyaan Shakti offers a favourable environment for setting up a business. Karma Shakti provides practical help to bussing women entrepreneurs to start their own business or scale them up.

vii. Venture Capital Scheme- It is a programme run by Small Farmers Agribusiness Consortium (SFAC) which aims to improve agricultural production. Under this programme, the SFAC helps in the development of the agricultural market and improvement in the marketing of agricultural products. The minimum value of the business should be Rs. 50 lakhs to become eligible for this programme.

viii. Standup India- The Government of India started the Standup India programme in 2016, with the sole aim of promoting entrepreneurship and startup cultures in India and is driven by the Small Industries Development Bank of India (SDBI). The SMEs can avail of loans of up to one crore under this programme. This scheme enables to avail a loan of up to 75 per cent of the total project cost.

4.12. Other Policy Initiatives for Skill development in India

1. In April 2016, MSDE along with NSDC and UK India Education Research Initiative (UKIERI) launched 'Transnational Skills Standards' in India, with the aim to support two major initiatives of the Government i.e. 'Make in

India' and 'Skill India' and to align to skill standards recognized globally. These standards in skill development are benchmarked to United Kingdom across 82 identified job roles. The aim is to benchmark the standards to support international mobility of Indian workforce and also prepare them to work in international companies having their operations in the country.

2. India International Skill Centre: MSDE under the Skill India Mission has set up India International Skill Centre (IISC) to provide skill training and certification benchmarked to international standards. In the pilot phase, IISCs were set up through the NSDC to implement two schemes, Pradhan Mantri Kaushal Vikas Yojana (PMKVY) and Pravasi Kaushal Vikas Yojana (PKVY) for youth seeking global mobility for jobs. As part of IISC Programme, both domain skill training on international standards and Pre-Departure Orientation Training (PDOT) are being imparted to candidates.
3. The National Education Policy (NEP) 2020 provides an overarching framework for exposing at least 50 per cent of all learners to vocational education by 2025 and its integration with formal education.

Indian Council of Agricultural Research (ICAR) through its Extension Division undertakes technology assessment, demonstration and capacity building of farmers. ICAR launched the ARYA (Attracting and Retaining Youth in Agriculture) project in 25 districts of 25 states to train youth in entrepreneurial activities. In each of these districts, 200-300 youth are being empowered to establish their income generating units and work as model for other youth in the villages. Agrinnovate India Ltd. was set up in October 2011 as a for-profit company owned by DARE, incorporated under the Companies Act. Its objective is to promote agricultural innovations and entrepreneurship by leveraging the strengths of ICAR and DARE at creating public private partnerships in the area of capacity building.

4.13.International Partnerships and Collaborations for Skill Development

MSDE of the Government of India has sought active collaboration with several countries like the UK, Australia, Germany, France, the US, and Singapore in the area of technology transfer, training of trainers, and designing model centres of excellence. The India-Israel collaboration is focused on improving water use efficiency and agricultural productivity leveraging the innovation and technology strength of Israel.

To enable skill development at the grassroots level through the right partnerships and leveraging existing knowledge with farmers, the German Federal Ministry for Economic Cooperation and Development (BMZ) under its Global Initiative on Promoting Green Innovation Centres under the One World, No Hunger Initiative (SEWOH) and the Gesellschaft für Internationale Zusammenarbeit (GIZ) are supporting green colleges in India. This initiative is focused on skilling rural tribal youth to empower them within their own communities by means of gainful employment.

India has been working extensively in several countries in Africa to strengthen the skill quotient of their agricultural workforce and professionals through bilateral collaboration as well as through co-operation with third countries. The India-US partnership in "Feed the Future: India Triangular Training Programme" is such a flagship partnership. Training programmes under this partnership are organised by MANAGE, Hyderabad.

MSDE through its International Cooperation Division, National Skill Development Corporation (NSDC), Directorate General of Training (DGT) and other implementing organizations in coordination with the Ministry of External Affairs (MEA) works to facilitate the supply of a skilled Indian workforce abroad through harmonization of skill qualifications, mutual recognition of certificates, accreditation of training providers, etc.

Bilateral Engagements and Collaborations on Skill Development

(a) **G2G MoUs:** MSDE has signed 11 G2G MoUs with countries like Australia, Denmark, France, Germany, Japan, Qatar, United Arab Emirates (UAE) and United Kingdom (UK) in the field of skill development and vocational education training.

(b) **B2B MoUs:** NSDC under the aegis of MSDE handles B2B MoUs which also facilitates recruitment, migration and training services for many countries. At present, NSDC has signed 18 B2B MoUs with countries like Japan, Australia, Germany, UAE, Malaysia, Canada, Kingdom of Saudi Arabia, (KSA) etc. for mobility of skilled and certified Indian workers to GCC countries in hospitality and MEP (mechanical, electrical and, plumbing) job roles and Indian workers' mobility to Australia in job roles like Truck Driver, Drill Fitter, Chef, Machinist, etc.

Country Specific Engagements

Australia

(a) **Agreement between NSDC and VETASSESS:** This MoU aims to cooperate on operationalizing opportunities for migration of skilled workforce from India under various Australian Industry Labour Agreements. A pilot project under Australia's Horticulture Industry Labour Agreement is initiated for identified job roles (Irrigationist, Production Horticulture Supervisor, Production horticulturist, Machinery manager and Mobile plant operator).

(b) **Agreement between NSDCI and Perdaman Global Services (PGS):** This Agreement aims to provide a conducive platform for "Promoting International Workforce Mobility (IWM) of Skilled and Certified Indian Nationals to Australia.

Germany

The present engagements with the Federal Ministry of Education and Research, Germany abbreviated as BMBF are:

1. Indo German Vocational Education Training (IGVET II) - to bring all the actors from public and private sectors work together to create conditions conducive to improving cooperative vocational training in India.
2. BMBF Project: BMBF has funded a complementary project on strengthening company models of dual VET in industrial clusters in Northern India through a bottom-up approach (SINADE). The aim of the project is to develop a corporate blueprint for dual vocational training that can be deployed to other clusters later on.
3. IGnITE (Indo-German Initiative for Technical Education) project – This skill development project is run by the GIZ and Siemens India, in coordination with DGT under the aegis of MSDE. This programme provides support to selected Government ITI's in five selected trades, Electrician, Electronics Mechanic, Fitter, Turner and Machinist.
4. BMBF Project QualIndia - BMBF is funding a research project called QualIndia (Quality Analysis in Indian Vocational Education and Training Institutions) with a focus on Industrial Training Institutions and Polytechnic Colleges. The basic objective of the project is to develop and provide a demand-oriented quality model that is explicitly adapted to the Indian context and will be operationalized and tested in practice.
5. iMOVE and German training providers activities- New Skilled Immigration Act in Germany (March 2020) is also focusing on skilled people with a vocational qualification and even on young people who are searching an apprenticeship in Germany (max. age is 25, B2 level in German language). German training providers are offering preparation and bridging course .(e.g. transport and logistics, hospitality and tourism, agriculture and crafts).

A Joint Declaration of Intent (JDoI) between MSDE and Federal Ministry of Education and Research (BMBF) and Federal Ministry of Economic Cooperation and Development (BMZ) was signed in November 2019 in New Delhi, focusing on cooperation in the field of Skill Development and VET, Dual apprenticeships and workplace-based skill development, etc.

Japan

Technical Intern Training Programme (TITP) commenced in 1993 in Japan, aimed to promote international cooperation for transferring professional skills to young and middle-aged youths of developing countries. TITP in India was officially launched from October 2017 as per the Memorandum of Cooperation (MoC) signed between the MSDE and the Ministry of Justice, the Ministry of

Foreign Affairs and the Ministry of Health, Labour and Welfare of Japan (Ministry of Japan).

A Basic Framework for 'Partnership for Proper Operation of the System Pertaining to Specified Skilled Worker' was signed between India and Japan in 2021. To address the shortage of labour in Japan, a new residency status called Specialised Skilled Worker (SSW) was introduced in Japan. It covers 12 sectors under which skilled workers from India who meet the skills requirements and qualify the Japanese language tests would be eligible for employment in Japan.

Gulf Cooperation Council (GCC) Countries

(a) **Training for Emirates Jobs and Skills (TEJAS):** To train, certify and place Indian workers in GCC Countries NSDCI launched TEJAS programme in 2022.

(b) In 2016, MSDE signed MoU with National Qualifications Authority (NQA), Govt. of UAE, to enhance skill development and mutual recognition of qualifications.

(c) MoU between National Qualifications Authority, Govt. of UAE and MSDE for benchmarking Qualifications, Assessment and Certification leading to incentivization by UAE Government for hiring skilled certified workers from India.

Singapore

Singapore Polytechnic (SP) with the support of grant from Temasek Foundation International (TFI) is collaborating with NSDC to strengthen the training of trainers and assessors program pertaining to short term skilling across India and develop Quality Assurance Frameworks for the performance monitoring of the Trainer and Assessor Academies.

MoU was signed between NSDC and National University of Singapore's Institute of System Sciences (NUS-ISS) in May 2018 to collaborate in adequately skilling India's workforce in emerging technologies such as Big Data Analytics, Artificial Intelligence and Robotics.

Finland

(a) An MoU was signed between NSDC and EDUFI (National Development Agency) during January, 2019 to strengthen TVET through the promotion and facilitation of knowledge exchange, new innovations and connecting providers from both sides within the vocational education and training sector.

(b) An MoU was signed between NSDC and Omnia (International commercial arm of four Finnish organizations with outstanding expertise in education, training and development cooperation) for capacity building workshops, development of entrepreneurship modules, joint certification courses for labour mobility, etc.

4.2 Virtual Global Skill Summit

To facilitate overseas mobility of the skilled workforce, the Ministry of External Affairs (MEA), Ministry of Commerce and Industry (MoCI), Ministry of Education (MoE) and Ministry of Skill Development and Entrepreneurship (MSDE) jointly organized the first Virtual Global Skill Summit (VGSS) during November 2022 with the Indian Ambassadors /High Commission representing India Missions from ten nations. The summit aimed at institutionalizing a robust mechanism for exchange of information on skill requirements of countries and the skill availability in India. The summit stressed the need for thorough sectoral and geographic mapping of the available skills in India.

Source: https://nationalskillsnetwork.in>global-summit

References

Ali Baig, I. and Safdar Ali, M 2019 The role of Government Policies and Programmes for Skill development in India: Issues and Challenges https://www.researchgate.net>publication>links

Das, P.K. and Kar, S 2016 Public Expenditure, Demography and Growth: Theory and Evidence from India. IZA Discussion Paper no 9721.

Ashoka, Mody and Shekhar Aiyar 2011. The Demographic Dividend: Evidence from the Indian States, IMF Worlang Paper.

DGET 2014 Demand Responsive Vocational Training, Directorate General of Employment and Training, Ministry of L & E, New Delhi

FICCI-KPMG 2012 Skilling India- A look back at the Progress, Challenges and the Way Forward: FICCI-Ernst & Young 2012 Knowledge Paper on Skill Development in India- Learner First

Govt of India 2015 Draft National Policy for Skill Development and Entrepreneurship, Ministry of Skill Development and Entrepreneurship, New Delhi

MSDE 2019 Schemes and Initiatives, https://www.msde.gov.in>schemes-initiatives

Niti Aayog 2020 Empowering migrant workers through skill development and livelihood generation: Best Practices for State and District, Govt of India, Nov 2020

Planning Commission, XII Five Year Plan: Employment and Skill Development

Sharma, L and Nagendra, A 2016 Skill Development in India: Challenges and Opportunities, Indian Journal of Science and Technology, Vol 9 (48): 1-8

Sharma, H and Mehrotra, S 2025 Skill Development in India: The facts behind the figures, Feb 25, 2025 https://www.idronline.org

CHAPTER 5

INDIAN SKILL STRUCTURE AND GOVERNANCE

Skill and Education is on the Concurrent list of the Constitution. While policy, process and standards are the responsibility of the Centre, implementation and training are state domains.

With one of the youngest populations, a median age of 28, India can harness its demographic dividend by nurturing a workforce equipped with employable skills and prepared for the needs of the industry. Though 65 per cent of India's fast-growing population is under 35, many of them lack the skills needed by a modern economy. However, it must be noted that the percentage of skilled workforce has improved from around 34 per cent to 51.3 per cent in the last decade. The Indian government has been actively working to enhance skilling and employment opportunities across the nation, recognizing the pivotal role of human capital in driving economic growth and innovation.

The recent launch of the Skill India Digital platform aimed at achieving skilling, education, employment, and entrepreneurship ecosystem marks a notable step towards the "ease of acquiring skill" in India. The rise in the number of candidates undergoing skill development through the Government's flagship programmes has underlined the thrust to 'Skill India'.

Under the Union Budget 2024-25, a highlight was the announcement of a new centrally sponsored scheme under the Prime Minister's package, in collaboration with state governments and industry. This scheme aims to skill 20 lakh youth over five years and upgrade 1,000 Industrial Training Institutes (ITIs).

Additionally, the announcement was made to revise the Model Skill Loan Scheme to facilitate loans up to Rs. 7.5 lakh with government-backed guarantees to benefit 25,000 students annually. For those ineligibles for existing schemes, financial support for loans up to Rs. 10 lakh for higher education in domestic institutions will be provided, with e-vouchers offering annual interest subvention of 3 per cent for one lakh students each year.

5.1 SIDH

The Skill India Digital Hub (SIDH) platform, launched in August 2023, represents a convergence platform facilitating access to skilling, credit, and employment through AI/ ML technology. This initiative integrates a comprehensive array of skilling schemes along with 690 online courses and 1650 QP-based e-books, enhancing accessibility to educational resources essential for vocational training. Furthermore, the platform seamlessly incorporates various government initiatives and services such as eShram/ EPFO/NCS, Udyam, DigiLocker, GatiShakti, UMANG, AgriStack, PLI Schemes, ODOP, etc.

Source: https://www.skillindiadigital.gov.in

The key achievements/ initiatives by the MSDE during 2024 are:

1. **Skill Loan Scheme:** MSDE launched the revised Model Skill Loan Scheme during July, 2024 with the objective of providing easy access to advanced-level skill courses, which potentially pose a significant financial barrier for many deserving students and candidates to gain futuristic and in-demand industry skills. Under the new Model Skill Loan Scheme, the maximum eligible loan amount has been increased from Rs. 1.5 lakh to Rs. 7.5 lakh.
2. **World Youth Skills Day**: In 2014, the UN General Assembly declared July 15 as World Youth Skills Day to celebrate the strategic importance of equipping young people with skills for employment, decent work and entrepreneurship. MSDE organised an open house, "Kaushal Samwaad" to celebrate World Youth Skills Day in 2024. The day also commemorated the tenth year celebrations of the Skill India Mission (launched on 15th July, 2015).

5.1.Skill Governance

Skill India Mission was launched in 2015, aimed to create and implement comprehensive skill development training programmes that would help bridge the gap between industry demands and skill requirements and thus, develop the country at large.

The Skill India programmes comprised implementing curriculum-based skill training courses, wherein trainees would gain certifications and endorsements from industry-recognised learning centres. The mission also involved incorporating skill-based learning in the school curriculum, creating opportunities for both long and short-term skill training and employment.

Under the Skill India Mission, the government established key institutions to govern and support various skill development programmes. They are: i) Ministry of Skill Development and Entrepreneurship (MSDE, ii) National Skill Development Corporation (NSDC) and iii) Sector Skill Councils (SSC).

The Ministry of Skill Development and Entrepreneurship (MSDE) has been created to fulfil the vision of a 'Skilled India' where human resource development is the primary focus. MSDE is responsible for co-ordination with all concerned for evolving an appropriate skill development framework, removal of disconnect between demand for, and supply of, skilled manpower, skill upgradation, building of new skills, innovative thinking and talents for existing and future jobs. MSDE also plays the lead role in ensuring the implementation of the National Policy for Skill development and Entrepreneurship 2015. The Ministry aims to skill on a large scale with speed and high standards in order to achieve the vision of a 'Skilled India'.

Under the Allocation of Business Rules, 1961, the Ministry of Skill Development and Entrepreneurship (Kaushal Vikas aur Udhyamita Mantralaya) is responsible for the following skill development efforts of the Central Government.

1. Skill Development Framework: Co-ordinating with relevant stakeholders to evolve an appropriate skill development framework. This includes bridging the gap between demand and supply of skilled manpower through vocational and technical training, skill upgradation, and fostering innovative thinking.
2. Skills Mapping and Certification: Mapping existing skills and ensuring their certification.
3. Youth Entrepreneurship Education: Expanding youth entrepreneurship education by forging strong partnerships between educational institutions, businesses, and community organizations. Setting national standards for entrepreneurship education.
4. Co-ordination for Skill Development: Ensuring effective co-ordination in skill development efforts.
5. Market Research and Training Curriculum: Conducting market research and devising training curricula for important sectors.
6. Industry-Institute Linkage: Facilitating collaboration between industries and educational institutions.
7. Public-Private Partnerships: Encouraging partnerships with industries requiring skilled manpower
8. Policy Formulation: Developing broad policies related to market requirements and skill development across other Ministries/Departments
9. Soft Skills Policies: Framing policies related to soft skills
10. IT and Computer Education: Promoting large-scale skill development in Information Technology and computer education.
11. Academic Equivalence: Ensuring equivalence of skill sets with academic qualifications.

12. Industrial Training Institutes (ITIs): Overseeing work related to ITIs
13. National Skill Development Organizations:
    - National Skill Development Corporation (NSDC)
    - National Skill Development Agency
    - National Skill Development Trust
14. Skilling for Entrepreneurship in Science and Technology
15. Institutes for Entrepreneurship and Small Business Development
    - National Institute for Entrepreneurship and Small Business Development (NOIDA)
    - Indian Institute of Entrepreneurship (Guwahati)

The MSDE is aided in various skill initiatives by its functional arms: Directorate General of Training (DGT), National Council for Vocational Education and Training (NCVET), National Skill Development Corporation (NSDC), National Skill Development Fund (NSDF) and 37 Sector Skill Councils (SSCs) as well as 33 National Skill Training Institutes (NSTIs), about 15000 Industrial Training Institutes (ITIs) under DGT and 187 training partners registered with NSDC. The Ministry also intends to work with the existing network of Skill Development centres, universities and other alliances in the field. Further, collaborations with relevant Central Ministries, State governments, international organizations, industry and NGOs have been initiated for multi-level engagement and more impactful implementation of Skill Development efforts.

MSDE's Vision 2025 adopts an ecosystem-enabling lens to transition India to a high-skills equilibrium and help create positive outcomes for individuals, enterprises and the economy. The three outcomes to be achieved through vision are:

- Enable individual economic gains and social mobility;
- Create a skills market that is learner-centric and demand-driven; and
- Facilitate aspirational employment and entrepreneurship generation, improve overall productivity for enterprises, and catalyse economic growth

5.2. Skill India Programme

Pradhan Mantri Kaushal Vikas Yojana 4.0 (PMKVY 4.0), the Pradhan Mantri National Apprenticeship Promotion Scheme (PM-NAPS) and the Jan Shikshan Sansthan (JSS) Scheme, the three key components, are now combined under the composite Central Sector Scheme of "Skill India Programme". These initiatives aim to provide structured skill development, on-the-job training, and community-based learning, ensuring that both urban and rural populations, including marginalized communities, have access to high-quality vocational education.

PMKVY 4.0 scheme provides NSQF aligned skill development training through Short-Term Training (STT) including Special Projects (SP) and reskilling and upskilling through Recognition of Prior Learning (RPL) with its target beneficiary being 15-59 years of age. The Pradhan Mantri Kaushal Vikas Yojana 4.0 (PMKVY 4.0) has undergone transformational changes to make skill development training industry-oriented, aligned with national priorities with increased accessibility.

Additionally, the scheme places a strong emphasis on international mobility, ensuring Indian workers are equipped with globally recognized skills. The Ministry has Mobility Partnership Agreements (MMPAs) and MoUs with various countries and has conducted necessary sectoral skill gap studies. Under the scheme, enablement of training in domain skills, joint certifications, language proficiency, and soft skills have been initiated to enhance the international mobility opportunities for our workforce.

Under PMKVY 4.0, a whole-of-government approach has been adopted to drive inter-ministerial convergence, ensuring the seamless execution of skilling initiatives across sectors. The scheme caters to the skilling components of various skill development and entrepreneurship schemes, maximizing impact and resource efficiency. The key collaborations include PM Vishwakarma under the Ministry of Micro, Small & Medium Enterprises, PM Surya Ghar: Muft Bijli Yojana, and the National Green Hydrogen Mission of the Ministry of New and Renewable Energy, NAL JAL Mitra, etc.

The Pradhan Mantri National Apprenticeship Promotion Scheme (PM-NAPS) supports seamless transition from education to work, ensuring apprentices gain industry-specific skills through real-world exposure. To support both apprentices and establishments in India, 25 per cent of the stipend, up to Rs.1,500 per month per apprentice, will be provided through Direct Benefit Transfer (DBT) during the training period, provided by the Central Government. The scheme is designed for individuals aged 14 to 35 years, ensuring inclusive access to skill development opportunities across various demographics.

The Jan Shikshan Sansthan (JSS) scheme is a community-centric skilling initiative designed to make vocational training accessible, flexible, and inclusive, particularly for women, rural youth, and economically disadvantaged groups and caters to the age group of 15 -45 years. By delivering low-cost, doorstep training with flexible schedules, JSS ensures that skilling opportunities reach those who need them the most, fostering both self-employment and wage-based livelihoods. Beyond skill development, the programme plays a vital role in social empowerment, creating awareness on health, hygiene, financial literacy, gender equality, and education within communities JSS is linked with key initiatives of the Government like: Pradhan Mantri Janjadi Adivasi Nyaya Maha AbhiyaN (PM JANMAN), Understanding of Lifelong Learning for All in Society (ULLAS), etc. to promote inclusive skilling.

Aligned with national frameworks, all certifications under the Skill India Programme are mapped to the National Skills Qualification Framework (NSQF) and seamlessly integrated with DigiLocker and the National Credit Framework

(NCrF), ensuring formal recognition of skills and enabling smooth transitions into employment and higher education.

Pradhan Mantri Virasat Ka Samvardhan (PM VIKAS), is a skilling initiative from Ministry of Minority Affairs (MoMA) focusing on the skilling, entrepreneurship and leadership training requirements of the minority and artisan communities across the country. The scheme is intended to be implemented in convergence with the 'Skill India Mission' of Ministry of Skill Development & Entrepreneurship and through integration with the Skill India Portal (SIP).

5.3.Training Providers in Skill Development

NSDC provides support to build scalable and profitable vocational training institutions across the country. It also acts as a catalyst in skill development by providing funding/technical assistance to enterprises, companies and organizations associated with skill development. NSDC has played a pivotal role in developing various types of partnership models, as below:

- Funded partnership: NSDC provides funding support for commercially viable, scalable and sustainable businesses in the skills domain, run by any registered legal entity.
- Non funded partnership: NSDC partners with reputed entities, not requiring financial assistance
- Innovation partners: NSDC identifies and provides support to extraordinary innovative enterprises.

With a Training Partner (TP) network of over 500 affiliates, 37 industry-led Sector Skill Councils (SSCs) and various enabling systems and initiatives, NSDC is a major contributor to the Skill India Mission.

On-boarding Process of Training Partners

After signing of official agreement between NSDC and a Training Partner, essential documents including copy of the agreement, term sheet, financial model and primary contact details of the TP has to be submitted to the Monitoring Team at NSDC.

The Monitoring team initiates the TP's onboarding process by inviting the entity for project kick-off meeting. The agenda of the meeting is to get the TPs acquainted with NSDC's ecosystem, monitoring processes, branding, database management, financial audits guidelines, and quality control mechanisms, registration in Skill India portal etc. Some of the sessions covered in the kick-off meeting include:

- Monitoring and Management Information System (MIS)
- NSDC Branding and Social Media
- Learning Resources and Technology

- Skill India Portal (SIP)
- Disbursement Process and Financial Reporting Requirements

After the meeting, key contact details of the training partner, with respect to the project are collected, including Senior Management, SPOC from Monitoring & MIS Department and Finance Department for further communication. Every TP is also allocated a SPOC (Single Point of Contact) from NSDC monitoring team, who acts as a facilitator and coordinator for all day-to-day communication, service related requests and issues/grievances management.

During the kick-off meeting, NSDC issues Partnership Certificates to the newly inducted training partners, both funded and non-funded as a proof of affiliation with NSDC. Only TPs with a valid Partnership Certificate and/or Agreement have the right to use NSDC Branding (as per the Branding Guidelines set by NSDC) and claim benefits of NSDC affiliation.

Directorate General of Training:

The Directorate General of Training (DGT) in the Ministry of Skill Development and Entrepreneurship is the apex organisation for development and coordination at National level for the programmes relating to vocational training including Women's Vocational Training. Industrial Training Institutes (ITI) are under the administrative and financial control of State Governments or Union Territory Administrations. DGT also operates Vocational Training Schemes in some of the specialized areas through field institutes under its direct control. The development of these programmes at national level, particularly in the area concerning common policies, common standards and procedures, training of instructors and trade testing are the responsibility of the DGT. Since DGT has been striving ahead in this technological age for hand holding the youth across the country, it has ensured digital skilling and industry readiness of the 23.15 lakh trainees in the ITI ecosystem by making 'Employability Skills' as a mandatory component of their curriculum.

The key players in the skilling ecosystem in India are the **Awarding Bodies, Assessment Agencies and Training Providers;** wherein the Awarding Body articulates job standards and certifies competencies and Assessment Agency follows an assessment and validation procedure that formally evaluates the learning outcomes (knowledge, skills and/or competencies) of an individual.

AICTE Affiliated organizations:

Skill Development Cell under AICTE has been entrusted the responsibility to train youth by providing them skills through AICTE approved Colleges/ Registered facilitators with objective for enhancing their Employment/ Self-Employment opportunities. The major schemes are: Pradhan Mantri Kaushal Vikas Yojana for Technical Institutes (PMKVY-TI), Employability Enhancement Training Programme (EETP), National Employability Enhancement Mission (NEEM), AICTE- Startup Policy, Skill Assessment Matrix for Vocational Advancement of Youth (SAMVAY).

PMKVY-TI: This scheme is implemented through AICTE approved Colleges to impart Engineering skills to drop-out students and find placement in suitable private sector jobs.

Employability Enhancement Training Programme (EETP): For implementation and enhancement of employment opportunities under skill initiatives, the Council has signed MoU's with LinkedIn, ICT Academy and Monster.com under EETP.

National Employability Enhancement Mission (NEEM): The objective of NEEM is to offer on the job practical training to enhance employability of a person either pursuing his or her graduation / diploma in any technical or non-technical stream or have discontinued studies of degree or diploma course to increase their employability

Start-Up Policy for Technical Institutions: AICTE has prepared a Start-up Policy for students of Technical Institutions to create tech-based student owned start-ups and employment opportunities.

As in all sectors of education, there are three broad group of training providers

- Government managed and funded organizations
- Privately managed and government funded (aided) organizations
- Privately funded and managed organizations

National Skill Development Corporation (NSDC), endeavours to create a sustainable and enabling skill ecosystem. In order to foster skill development across the country, NSDC provides funding support to enterprises, companies and organizations to build capacity by extending resources, technical assistance, thought leadership and knowledge management. Playing the role of a market-maker, NSDC is responsible for enhancing and strengthening the skill development initiatives and ensuring long-term stability and spurring growth.

NSDC provides concessional loans (soft loans) to Training Partners, covering up to 85 per cent of the total project investment (remaining portion is invested in the form of equity infused by the TP) to cover the expenditure related to skill training.

The Ministry of Social Justice and Empowerment, Government of India, has launched the Pradhan Mantri Dakshta Aur Kushalta Sampann Hitgrahi (PM-DAKSH) Yojana, a Central Sector Scheme, from 2020-21, for providing for trainings to marginalized persons covering, inter-alia, the Scheduled Castes persons between the age of 18 to 45 years.

The main objective of PM-DAKSH Yojana is to develop the skills of target groups and make them employable either in self-employment or wage-employment. The rural artisans belonging to the target group, who have become marginalized due to the presence of better technologies in the market, would also be trained so as to enable them adopt better process and increase their incomes for improved living.

The National Scheduled Castes Finance and Development Corporation (NSFDC) under Ministry of Social Justice and Empowerment (MSJE) has been sponsoring employment oriented NSQF compliant Skill Development Training Programmes (SDTP) for the persons belonging to Scheduled Castes. In case of SDTPs, there is no income criterion for the aspirants. The Skill Development Training Programmes are being implemented through the Training Institutes selected by MSJE, on year-to-year basis.

There are four types of Skill Development Training Programmes under PM-DAKSH Yojana:

1. Up-skilling/Re-skilling - Normally, having duration from 35 to 60 hours and spaced over up to 35 days.
2. Short Term Training - Normally, having duration from 200 to 300 hours and up to 3 months, as stipulated in National Occupational Standards (NOS) and Qualification Packs (QPs).
3. Entrepreneurship Development Programme (EDP) – Normally, having duration up to 90 hours/15 days or as stipulated by Ministry of Rural Development, Government of India.
4. Long Term Training - Normally, having duration up to 650 hours/7 months, as stipulated by the concerned Board/Regulatory Body of the Training Centre.

Under the programmes, the trainees are provided free training and stipend @ Rs.1500/- per month after the training, subject to overall attendance percentage of a trainee is 80% and above in the entire training period, through Direct Benefit Transfer (DBT).

On successful completion of training and assessment & certification of trainees, they are provided placement assistance and/or entrepreneurial guidance to start their own ventures with financial assistance from NSFDC through State Channelizing Agencies/ Channel Partners.

The Skill Development Training Programmes under PM-DAKSH Yojana, are being implemented through dedicated PM-DAKSH Portal.

5.4. Certification Bodies

Certification bodies play a crucial role in skill development in India. Some key organizations involved in certification are: NSDC, Regional Skill Development and Certification Authority (RSDCA), Sectoral Skill Councils and World Skill Council.

RSDCA is an autonomous body established Indian Trust Act with the prime vision and objective to promote and develop the Minority Community who are unable to undergo education due to their financial constraints and being placed in the remote area. RSDCA authorised training centres attempts to bridge this widening divide between those who have access to opportunities and those who are increasingly being marginalized from the 'new economy' jobs. RSDCA addresses this endemic problem with a market-based approach that is sensitive to the socio-economic needs of marginalized youth in the 16-45 years age group.

The Certification Bodies are categorised as:

1. Government backed certification bodies
    - NSDC
    - DGT
    - NCVT
2. Sector specific certification bodies
    - Agriculture Skill Council of India (ASCI)
    - Healthcare Sector Skill Council (HSSC)
    - It-ITes Sector Skill Council (NASSCOM)
3. International Certification bodies
    - City and Guilds: A UK based certification body that offers vocational training and certification programme in India
    - Pearson Vue: A global certification body that offers certification programmes in various fields including IT, healthcare and finance
    - Certified Associate in Project Management (CAPM) offered by PMI
4. Other Certification bodies
    - Indian Institute of Skill Development
    - Skill Council for Green Jobs (SCGY)
    - Media and Entertainment Skill Council (MESC)
5. Certification bodies in Agriculture sector
    - IFOAM
    - International Organic Accreditation Service (IOAS)
    - USDA- NOP
    - EU Organic
    - Japan Agricultural Standards (JAS)

5.5. Sector Skill Councils

With an aim to strengthen the demand supply linkage in skill development and step up the level of private industry (potential employers) engagement in skill development programmes, the NSDM included the concept of Sector Skill Council (SSC) in 2015.

SSCs are responsible for identifying skill development needs per sector, listing out the details for candidates to choose from and maintaining a repository of skills. SSCs are involved in creating skill development plans, in determining skill/competency standards and qualifications, and in notifying the same as per the NSQF. They can also conduct skill-based assessment and certification of Qualification Packs (QPs) and National Occupational Standards (NOS).

SSCs are also tasked with promoting Centres of Excellence (CoE) and conducting train-the-trainer programs together with the states and the NSDC.

Among the 37 Sector Skill Councils established by the NSDC, the Agricultural Sector Skill Council of India (ASCI) and the Food Industry Capacity and Skill Initiative (FICSI) cater to the skill development requirements of the agriculture and food sectors. (Annexure-3)

Agricultural Skill Council of India (ASCI)

Agriculture Skill Council of India (ASCI) is a Section 8 Not for profit concern working under the aegis of MSDE. ASCI works towards capacity building by bridging gaps and upgrading skills of farmers, wage workers, self-employed and extension workers engaged in organized / unorganized segments of Agriculture and allied sectors.

ASCI is contributing to nation building through Skill Development in Agriculture especially at a time when country's agriculture is experiencing stagnant growth, exodus of quality manpower to other sectors, changing climate with increased variability in production parameters and transformations in international agriculture markets, especially too much subsidies challenging the competitiveness of Indian Agriculture.

ASCI has taken upon itself the responsibility of transforming Indian Agriculture through developing the skills of country's manpower in emerging areas of agriculture. With the development of 176 Qualification Packs, ASCI has covered segments like Farm Mechanization, Dairy Farm Management, Seed Industry, Poultry Farm Management, Fisheries, Soil Health Management, Post-Harvest Supply Chain Management, etc.

ASCI works towards capacity building by bridging gaps and upgrading skills of farmers, wage workers, self-employed and extension workers engaged in organized / unorganized segments of Agriculture and allied Sectors.

ASCI has identified the emerging areas for Skill Development in Organic Farming, Hi-tech Horticulture, Precision Farming, Protected Cultivation, Animal Husbandry, Fisheries, Poultry, Micro-Propagation, Integrated Natural Resource Management, Integrated Pest & Nutrients Management, Production of Quality Seeds, Watershed Management, Post-harvest Management & Value Addition, Application of Biotechnology, Information and Communication Technologies, Geographic Information System (GIS) mapping and Conservation of biodiversity of plants, animals, fish, and microbes.

NCVET has recognised Agriculture Skill Council of India as its Awarding Body.

The ASCI is also working with the Ministry of Human Resource Development through Rashtriya Madhyamik Siksha Abhiyan (RMSA), an agriculture skill development programme aimed at students from grades 9 (equivalent to level 01 of the NSQF) and 12 (level 04). In addition to the support extended to the state education board, the ASCI provides QPs for training, and undertakes assessments and trainings.

5.2 Job Roles in Fisheries

In the changing scenario of fisheries and aquaculture, ASCI has identified 26 job roles in consultation with industry and fishery associations.

In aquaculture, the job roles are: Aquaculture Technician, Aquaculture Worker, Freshwater Aquaculture Farmer, Hatchery Manager, Hatchery Product Worker, Fish seed grower, Brackish water Aquaculture farmer, Shrimp farmer, Feed Technician, Aquarium Technician, Mariculture Operator, etc.

In capture fisheries the job roles are: Marine fisherman, Inland fisherman, Fishing boat driver, Fishing boat deckhand, Fishing boat maintenance worker, Fishing boat mechanic, etc.

Recently, two more job roles are added- Deep sea fisherman and Seaweed cultivator

The Food Industry Capacity & Skill Initiative (FICSI) is the Food Processing Sector Skill Council established as a non-profit organization, registered under the Societies Registration Act, 1860. The FICSI is being promoted by the Federation of Indian Chambers of Commerce and Industry (FICCI) and financially supported by the NSDC. Like other SSCs, its mandate is to generate a pool of skilled manpower to meet the skill requirements of the food processing industry. The FICSI has developed 45 QPs cutting across various operations related to the food processing industry.

The Ministry of Food Processing Industries (MoFPI) is working with the FICSI to complete the validations of QPs for each job role. It is also involved in the development of course curricula through the National Institute of Food Technology Entrepreneurship and Management (NIFTEM). Under Pradhan Mantri Kisan SAMPADA Yojana, MoFPI supports skilling infrastructure and development of course curricula.

5.6. Assessment Bodies

Assessment Agency is defined as an agency which tests or conducts examinations to assess whether a learner has met the requirements necessary to be competent and qualified with respect to a skill or qualification. As an overarching regulator, NCVET strives to establish a credible and quality assured assessment ecosystem through standardized norms, continuous quality governance and incentivization of better performing agencies, assessors/proctors. The vision is to create a pool of best-in-class Assessment Agencies for assessment of skill acquisition.

Skills assessment in India is a rapidly evolving sector. Assessment can verify that individuals have the skills needed to perform a particular job and that the learning programme undertaken has delivered education at a given standard. It enables individuals to be benchmarked against their peers. It also has more intrinsic value in creating opportunities to motivate students and to give feedback. Assessment is closely linked to certification, as individuals often have the opportunity to gain a recognised certificate if they pass an assessment; certificates can provide a platform from which to progress to other areas of learning, to get a job or to progress within their existing careers.

5.7.National Occupational Standards (NOS) & Micro Credentials (MC)

NOS guidelines specify the standard of performance an individual must achieve when carrying out a function in the workplace, together with the knowledge and understanding they need to meet that standard consistently. Each NOS defines one key function collating the related performance criteria in relation to the specific function in the job role concerned. With the advent of new policies like recognition of multiple awarding bodies, a need was felt to provide more flexibility in design of the qualifications. NOS approval and alignment was mandated to establish and facilitate:

- Cross Sectoral and Multi Sectoral Qualifications design
- Use of available knowledge
- Up Skilling for allowing opportunity to choose unit of competencies as per requirement
- Bridge Courses and RPL with upskilling
- Lifelong Learning
- Multiple Entry & Multiple Exit (ME-ME) Pathways
- Better channelization of resources

A micro-credential certifies the achievement of a coherent set of skills and knowledge and is specified by a statement of purpose, learning outcomes, and strong evidence of need by industry, employers, government, or the community. They are smaller than traditional qualifications and focus on skill development opportunities not currently catered to in the regulated skilling ecosystem.

Unlike traditional degrees and certificates, micro-credentials are generally offered in shorter or more flexible time spans and tend to be more narrowly focused Micro-credentials may be composed of units (performance criteria centred around identified learning outcomes) that can be developed within a sector or across sectors, typically in multiples of 7.5 learning hours per learning day. These can be structured as a set of four such days (30 hours), two such days (15 learning hours), or one such day (7.5 hours) with a Fast Track assessment conducted using online tools/modes However, based on requirements, the number of hours may vary.

5.8.National Credit Framework

To fulfil the vision of National Education Policy 2020, of making education more holistic and effective and integration of general (academic) education and vocational education, the Government of India had approved the constitution of a High-Level Committee in November, 2021, to develop a National Credit Accumulation & Transfer Framework for both Vocational and General Education.

The National Credit Framework is expected to fulfil the following objectives:

- Remove hard separation between General and Vocational education
- Establish equivalence between General and Vocational education and training/ skilling
- Enable mobility between and within General and Vocational education and training/ skilling
- Enable provisions for Lifelong Learning through Multiple Entry and Exit Options
- Integrate Education, Skilling, and Work Experience

The following initiatives are implemented to further boost skill development programmes in India:

- Pradhan Mantri Kaushal Kendras (PMKK) is extended to more than 600 districts across the country. 100 India International Skills Centres will be established across the country.
- Skill Acquisition and Knowledge Awareness for Livelihood Promotion programme (SANKALP) is launched at a cost of Rs 40 billion. SANKALP will provide market relevant training to 35 million youth.

5.9. Stages in Skill Value Chain

'Skill Value Chain' refers to the sequence of activities or processes involved in developing and leveraging skills within an organization or an individual. It represents the journey from identifying the skills needed to achieve desired outcomes to effectively utilizing those skills to create value. The Skill Value Chain encompasses various stages that interconnect and build upon each other. The stages are:

Skill Identification: The first stage involves identifying the skills required to meet organizational goals or individual objectives. It entails conducting a skills gap analysis to determine the current skill set and identify areas that need improvement.

Skill Acquisition: After required skills are identified, the next step is to acquire or develop those skills. This stage involves various methods such as training programmes, workshops, online courses, mentoring, on-the-job learning, or formal education.

Skill Application: After acquiring the necessary skills, this stage involves integrating the newly acquired skills into daily work or problem-solving situations. It includes activities like projects, simulations, case studies, or real-world experiences.

Skill Refinement: Continuous improvement is a vital aspect of the Skill Value Chain that ensures that skills remain up-to-date and aligned with changing needs. This stage emphasizes refining and enhancing skills over time through practice, feedback, reflection, and further learning opportunities.

Skill Utilization: The value of skills is realized when they are effectively used to accomplish specific tasks, solve problems, or contribute to organizational success. This stage involves deploying skills in a manner that adds value, enhances productivity, and achieves desired outcomes.

Skill Evaluation: Regular assessment and evaluation are necessary to gauge the effectiveness and impact of skills. This stage includes measuring skill performance, identifying strengths and weaknesses, and gathering feedback from various stakeholders.

Skill Alignment: The final stage of the Skill Value Chain focuses on aligning individual or organizational skills with strategic goals and future needs. This involves identifying emerging skill requirements, forecasting trends, and adapting the skill portfolio accordingly.

By understanding and actively managing each stage of the Skill Value Chain, organizations and individuals can optimize the development, application and utilization of skills, leading to improved performance and competitive advantage.

Skill Value chain framework

In the context of skill development initiatives in India, the Skill Value Chain refers to the framework and processes involved in enhancing the skills of individuals and bridging the gap between industry requirements and the available workforce. India has been focusing on skill development to equip its population with the necessary capabilities to meet the demands of a rapidly evolving economy. The Skill Value Chain applies to skill development initiatives in India, as below:

Skill Identification and Mapping: The process begins with identifying the skills required by various industries and sectors. This involves analysing market trends, conducting surveys, consulting with industry experts, and collaborating with employers to understand their skill needs. Skill mapping helps in aligning the training programmes with industry requirements.

Designing Skill Development Programmes: Once the skills are identified, the government, along with various training organizations and agencies, designs and implements skill development programmes. These programmes encompass a wide range of sectors, including manufacturing, agriculture, information technology, healthcare, construction, hospitality, etc. Skill trainings are provided through vocational institutes, ITIs , Skill Development Centres, and online platforms.

Training and Capacity Building: Skill development initiatives in India emphasize imparting training and building the capacity of individuals to acquire new skills. This includes both technical and soft skills. Training programmes are designed to be practical and industry-relevant, incorporating hands-on learning, simulations, and apprenticeships. They also include components like entrepreneurship development, communication skills, and digital literacy.

Certification and Recognition: Recognizing the importance of certifications, skill development initiatives in India focus on providing standardized certification for acquired skills. NSDC and other agencies collaborate with industry bodies to develop competency-based assessment frameworks and certification programmes. These certifications validate the skills acquired by individuals and enhance their employability.

Industry Collaboration and Placement: To bridge the gap between industry demand and skilled workforce supply, skill development initiatives foster strong collaboration with industry stakeholders. Public-private partnerships are formed to ensure industry participation in curriculum design, training delivery, and assessment. Placement cells and job fairs are organized to facilitate the placement of skilled individuals in relevant job roles.

Continuous Learning and Upgradation: Skill development is an ongoing process. Initiatives in India promote lifelong learning and upgradation of skills to keep pace with technological advancements and changing industry needs. They encourage individuals to engage in upskilling and reskilling programmes and provide access to platforms for continuous learning, such as e-learning portals and online courses.

Monitoring and Evaluation: Skill development initiatives in India employ monitoring and evaluation mechanisms to assess the effectiveness and impact of the programs. This involves tracking the progress of trainees, evaluating their skill development outcomes, and obtaining feedback from employers. Regular reviews and data analysis help in identifying areas of improvement and enhancing the quality of skill development initiatives.

By following the Skill Value Chain in skill development initiatives, India aims to create a skilled workforce that meets the demands of various industries, promotes inclusive economic growth, reduces unemployment, and enhances the country's global competitiveness.

Technical and Vocational Education Training or TVET activities can be classified systematically in the form of skills value chain. The value chain is intrinsically mapped to a student's life cycle from mobilization to training followed by assessment, certification and employment opportunities.

In a training ecosystem that NSDC has catalyzed in India, the TP is a registered organization with its main activity as training (as a not for profit or for profit company/society etc.) and it opens up various training centres, which are both mobile and stationary in nature. Training centres are either self-owned or based on franchisee arrangements.

5.10.Centres of Excellence (CoE) in skill training

Centres of Excellence are places where the highest standards are maintained in respect to skill development in a specific sector(s). CoEs for skilling include structures, entities that leverage shared resources from the private sector, government and public sector. By bringing the various stakeholders together, CoEs can support adoption of best practices, superior leadership, research and high quality training by focusing on a specific sector. (Annexure-5)

Some of the key features of CoEs include:

- Delivery of high quality, specialized and advanced skill training that caters to a niche market, where volumes are low making it less attractive to conventional TPs
- Demand-driven training that is achieved by close inter-dependency with the government and private sector
- Alignment of training activities with government initiatives for economic development, providing courses in high end fields, including training of trainers and assessors
- Potential to incorporate global standards due to involvement of international industry employers
- Undertaking of research, innovation and applied learning by co-locating with research and development facilities
- Focus on job roles involving advanced technical, engineering roles, industries with high need of workers skilled in technical aspects and manufacturing and scientific fields where R&D is critical.

The Centres of Excellence established for skill development in the farm sector are:

Centre of Excellence for Dairy Skills in India (CEDSI) established as an autonomous institution committed to dairy skill development in India is the first Centre of Excellence established by ASCI. CEDSI aims to ensure sustainability and profitability in the dairy sector through skilling and capacity building, policy advocacy, knowledge management and research. CEDSI works with stakeholders across dairy ecosystem including dairy farmers, wage workers, dairy cooperatives and corporates, research, academia, and central and state governments. It strives to be positioned as an innovator and leader in the dairy skilling space across the globe, who undertake and facilitates applied manpower research in dairy sector, capacity development in the areas of skilling and provides policy inputs at the national and state level to strengthen skill upgradation.

Centre of Excellence for Horticulture Skills in India (CEHSI)- Working under Centre of Excellence for Agricultural Skills in India (CEASI), CEHSI aims to enhance sustainability and profitability in the horticulture sector through skills development, capacity building, policy advocacy, knowledge management and research. CEHSI collaborates with various stakeholders within the horticulture ecosystem, including horticulture farmers, wage workers, horticulture corporates, research institutions, academics and both central and state governments. It undertakes and facilitates applied manpower research in the horticulture sector, fosters capacity development in various skill areas and provides valuable policy inputs at the national and state levels to strengthen the skill upgradation efforts.

Centre of Excellence for Farm Mechanization Skills in India (CEFMSI)- This Centre is dedicated to farm mechanization skills development in India. The Centre aims ensure sustainability and profitability in the farm mechanization sector, through skilling, capacity building and policy advocacy, knowledge management and research. It aspires to be recognized as an innovative and global leader in the domain of farm mechanization.

5.11. National Skill Universities

The National Education Policy (NEP) 2020 places significant emphasis on skilling and the integration of vocational education with mainstream education. It allows for multiple entry and exit points with appropriate certification. Skill Universities play a unique and crucial role in fulfilling the NEP 2020 mandate.

The educational landscape of India is undergoing a transformative shift, with a growing thrust on skill based learning to meet the rapidly evolving demands of the job market. The establishment of National Skill Universities is a testament to this paradigm shift. (Annexure-1)

Skill Universities are primarily regulated by the UGC under the UGC Act and Regulations. They are designed to focus on VETS, emphasising practical skills and industry relations. They are mandated to offer at least 50 per cent of their programme as skill based. Within these programmes, up to 70 per cent need to be dedicated to the skilling component. Skill Universities are focused on integrating VETS into higher education by developing higher order skills-based qualifications and research initiatives. Presently, there are 13 skill universities established in India. (Annexure-2)

5.12. Private Sector Engagement in Skill Development and Entrepreneurship

Skill development in India has been primarily led by the government with the private sector playing its roles in limited capacity. Public programmes on skill development have the scope for private sector participation in partnership and such models have been in progress. With the shift in focus of the skill development programmes from mere vocational training to innovation and entrepreneurship, there is a greater role that the private sector can play. Also, most importantly, the policy emphasis on aligning skill development with demand requires the participation of private players. This is particularly relevant as the industry is a potential employer and is also better versed with the changing technology and operations which need to be considered in the development of training modules.

The private players in the skill development space can be categorized into private companies who undertake skill development as part of their core business as well as CSR and/or through foundations and trusts, which work on larger social and human developmental issues. There are developmental agencies active in capacity building and skill development of targeted beneficiaries in terms of funding such programmes run by the government or other private players. In addition, there are private training institutes which conduct trainings for the public run programmes on skill development.

Skill development at the village level is also undertaken by self-help groups and/or farmer clubs, etc. Most of these players work in partnership with each other and also in partnership with the government.

Indian Society for Training and Development is a national professional body actively involved in the domain of capacity building of various stakeholders.

5.3 ISTD

The Indian Society for Training & Development (ISTD), established in April 1970, is a national level professional and non-profit society with a large membership of individuals and institutions involved in the domain of training and development of Human Resources from Government, Public and Private Sector Organizations and Enterprises; Educational and Training Institutions and such other Professional Bodies. The Society is affiliated to the International Federation of Training and Development Organizations (IFTDO), Geneva.

ISTD, based on its availability of trained manpower and spread of Chapters throughout the country had undertaken Pilot Project for assessment of candidates and on successful completion, has been appointed as the nodal agency for assessments for testing and certification of skills for all sectors in India.

Source: https://www.istd.in

As private agri-business players try to secure and expand existing markets through value added services, their required labour resources increase and the skill sets needed are also quite diverse. For example, in agriculture, precision farming which involves hi-tech mechanization will require people with advanced skills for developing these technologies, customizing them to meet local needs, operating the same on the farmers' fields as well as educating and training the farmers for the future.

Private companies such as Jubilant Bhartiya Foundation, ICICI Foundation, Mahindra Tech Foundation, Unilever India, and Hindustan Coca Cola Beverages Pvt. Ltd. are also contributing towards skill development through their CSR models.

Innovation and entrepreneurship are two emerging elements of the skill development agenda. Several private sector players including corporate giants like Google, Microsoft, GE Capital, and Tata Consultancy Services (TCS) have set up accelerators and incubators to nurture innovations and boost entrepreneurship. Financiers like venture capitalists and angel investors have come forward in a big way in financing start-ups critical for sustaining the venture. There has been a flurry of agriculture and food related start-ups and early stage companies in India thereby giving rise to a new generation of entrepreneurs.

Other private institutions involved in skill development activities include the Indian Society of Agribusiness Professionals (ISAP) which runs a skill development programme targeting unemployed agriculture and allied sector graduates across 12 states in India including Jammu & Kashmir, Punjab, Himachal Pradesh, Haryana, Uttarakhand, Rajasthan, Gujarat, Madhya Pradesh, Jharkhand, West Bengal, and Chhattisgarh.

5.4 ISAP

Indian Society of Agribusiness Professionals (ISAP) is a non-government, non-profit organisation working since 2001 for the empowerment of rural India through sustainable farm and non-farm livelihood development.

The mission of ISAP is to 'enhance the livelihood opportunities' in rural India 'through sustainable intensification of agriculture, skill development and market integration' in the field. ISAP activities had spread to 15 out of India's 29 states, integrating more than a million farm households into the agriculture marketing system, through a network of producer organizations.

Agribusiness in India consisted of five main types of activities:

(a) agricultural inputs which included seeds, fertilizers, pesticides, insecticides, livestock feed, organic manure and compost, planting material, electricity and diesel, repair and maintenance services;

(b) agricultural production which included fruits and vegetables, pulses, edible seeds, nuts and oils, grain milling products, dairy products, meat, fish and poultry, beverages;

(c) agricultural processing which included dehydration, canning, packaging, processing of herbal and medicinal plants, dairy products, poultry products, livestock, food processing;

(d) creating and maintaining agribusiness infrastructure like irrigation facilities, cold storage and warehousing, specialized packaging and transport, agri-clinics and service centres; and also harvesters, tractors, tools, implements and farm machinery; and

(e) trading, management and marketing, which included contract farming, soil mapping, weather sensing, retailing, supply chain management, technical training and seminars, consulting and capacity building, financing and insurance.

The Skill Development Vertical of ISAP is a dedicated wing focused on empowering individuals through vocational and professional training. Recognizing that skill development is the backbone of growth, this division aims to enhance employability and entrepreneurial ability among marginalized communities. As part of its skill development initiatives, ISAP has done capacity building and training of farmers in various govt projects and CSR projects. The skill development programs have been run under the principal of Strengthening Training, Building Sustainable Employability, Tracking Retention and Consolidating Demand Driven Geographies.

Source: https://isapindia.org

The Entrepreneurship Development Programme (EDP) is part of the Agri Clinic and Agri Business Centres (ACABC) scheme steered by the National Institute of Agricultural Extension Management (MANAGE). (Annexure-4)

The penetration of technology in remote rural areas is continuously throwing up new challenges which need to be addressed in a timely manner. Hence, skilled people are required to work closer to the farmers' fields and/or connected over a virtual platform. While people no longer want to continue in agriculture and particularly in farming, technology driven agriculture is opening up new employment opportunities for youth in both on- and off-farm jobs. Development agencies play a critical role in integrating skill development with social and human development, and for incorporating lessons learnt and best practices from across the world for an improved design, implementation and monitoring of these programs.

Some of the major agribusiness companies engaged in skill development are Mahindra and Mahindra through their Samriddhi and Tringo (rental model for tractors) initiatives, Tata Kisan Centres (TKS), TAFE, John Deere, and Syngenta.

Funding

National Skill Development Fund (NSDF) has been set up by Government of India with the objective of encouraging skill development in the country. A public Trust set up by Government of India is the custodian of the Fund. The Fund acts as a receptacle for all donations, contribution in cash or kind from all contributors (including Government, multilateral organizations, corporations etc) for furtherance of the objectives of the Fund.

NSDF meets its objectives through NSDC set up for building skill development capacity and forging strong linkages with the market. NSDC acts as a catalyst in skill development by providing funding to enterprises, companies and organizations that provide skill training. It also develops appropriate models to enhance, support and coordinate private sector initiatives.

5.13. National Skill Qualification Framework (NSQF)

NSQF is a nationally integrated education and competency-based framework that enables persons to acquire desired competency levels. The NSQF organizes qualifications according to a series of levels of knowledge, skills and aptitude. These levels, graded from one to ten, are defined in terms of learning outcomes which the learner must possess regardless of whether they were acquired through formal, non-formal or informal learning. It is, therefore, a nationally integrated education and competency-based skill and quality assurance framework that will provide for multiple pathways, horizontal as well as vertical. It includes vocational education, vocational training, general education and technical education, linking one level of learning to another higher level. This will enable a person to acquire desired competency levels, transit to the job market and, at an opportune time, return for acquiring additional skills to further upgrade their competencies.

The key elements of the NSQF provide:

(a) national principles for recognising skill proficiency and competencies at different levels leading to international equivalency

(b) multiple entry and exit between vocational education, skill training, general education, technical education and job markets

(c) progression pathways defined within skill qualification framework

(d) opportunities to promote lifelong learning and skill development

(e) partnership with industry/employers

(f) transparent, accountable and credible mechanism for skill development across various sectors

(g) increased potential for recognition of prior learning.

The qualification framework is beneficial to schools, vocational education and training providers, higher education institutes, accrediting authorities as well as industry and its representative bodies, unions, professional associations and licensing authorities. The biggest beneficiaries of such a framework are the learners who can judge the relative value of a qualification at a particular level on the framework and make informed decisions about their career progression paths.

Need for Qualification Framework in India

In India, general education and vocational education and training have been operating as separate verticals, with very little interaction between the two. This has resulted in hesitation amongst the youth in opting for vocational education and training as it is presumed that the concerned individual will be unable to acquire higher degrees and qualifications. In order to facilitate mobility from vocational to general education, and vice-versa, a qualification framework for India, the National Skill Qualification Framework (NSQF) will help make qualifications more understandable and transparent.

The need for the NSQF arises due to the following additional reasons:

(a) Till now the focus of education and training has been almost entirely on inputs. The NSQF is based on an outcomes-based approach, and each level in the NSQF is defined and described in terms of competency levels that would need to be achieved. Job roles corresponding to each of these competency levels would be ascertained with the involvement of industry, through the respective Sector Skill Councils (SSCs).

(b) The pathways of learning and progression, especially on the vocational education and training front, are either absent or lack clarity. There is no clear provision for vertical or horizontal mobility. The NSQF will make the progression pathways transparent so that institutes, students and employers are made clear as to what they can or cannot do after pursuing a particular course and address the issues of inequity and disparity in qualifications.

(c) There is lack of uniformity in the outcomes associated with different qualifications across institutions, each with its own duration, curriculum, entry requirements as well as title. This often leads to problems in establishing equivalence of certificates/diplomas/degrees in different parts of the country, which in turn impacts the employability and mobility of students.

(d) The negative perception associated with vocational education and training can be significantly removed by the development of quality qualifications that also permit acquisition of higher qualifications, including degrees and doctorates.

(e) There exists a large section of people who have acquired skills in the informal sector but who do not have the necessary formal certifications to attest to their skills. As a competency-based and outcomes based qualification framework, NSQF will facilitate Recognition of Prior Learning (RPL) that is largely lacking in the present education and training scenario

(f) Majority of Indian qualifications are not recognized internationally and vice-versa. This creates a problem for the students and workers as their international mobility is adversely affected and they often have to undergo a course again to get a qualification that is recognized in the host country. The NSQF will also help alignment of Indian qualifications to international qualifications in accordance with relevant bilateral and multilateral agreements. Many countries are already in the process of aligning their qualifications to international qualifications through qualification frameworks

(g) The credit accumulation and transfer system that will be integrated in the NSQF will allow people to move between education, vocational training and work at different stages in their lives according to their needs and convenience. It will be possible for a student to leave education domain, get some practical experience in industry and return to studies to gain qualifications to progress higher in his chosen career

International Experience with Qualification Framework

A paradigm shift from education based on inputs to education based on learning outcomes is taking place. Outcomes-based learning is a widely used term. The shift to learning outcomes is important for a number of reasons:

(a) It shifts focus from providers to users of education and training.

(b) By explaining what a learner is expected to know, understand or be able to do at the end of a learning process, individuals are better able to see what is offered in a particular course and how this links with other courses and programs.

(c) It increases transparency and strengthens accountability of qualifications – for the benefit of individual learners and employers.

The vast majority of the world's industrialized and transition countries are reforming their qualifications. At the same time, they are developing frameworks to relate these qualifications to each other and to reflect new demands in society and the labour market. The development of these systems is often linked to changes in higher education, technical and vocational education and training (TVET) and lifelong learning.

Many countries worldwide are in the process of introducing qualification frameworks. Though the theoretical principles of all frameworks remain largely similar, the objectives of launching the frameworks vary. Whether the emphasis is on increasing the relevance and flexibility of education and training programs, easing recognition of prior learning, enhancing lifelong learning, improving the transparency of qualification systems, creating possibilities for credit accumulation and transfer, or developing quality assurance systems, Governments are increasingly turning to qualifications frameworks as a policy tool for reform. In some cases, national developments are propelled by the emergence of regional frameworks (such as the European Qualification Framework). In many cases the implementation of qualification frameworks has been widely supported by international organizations and is often linked to aid money and even loans. There is increasing activity from international agencies in the area of qualifications frameworks: the Organisation for Economic Cooperation and Development (OECD), the International Labour Office (ILO), the World Bank (WB) and the European Union (EU) have current qualification framework projects.

NEAT (National Educational Alliance for Technology)

Ministry of Education, GoI has announced a National Educational Alliance for Technology (NEAT) as a Public-Private partnership model between the Government (through its implementing agency AICTE) and the Education Technology companies of India. Through an open invitation and screening, companies are invited to showcase their products on a National Portal developed for the learners, who may procure them based on their requirements.

The aim of NEAT is to bring the best technological Products in education pedagogy on a single platform for the convenience of learners. Technology Products using Artificial Intelligence for customized learning or e-content in niche areas having highly employable skills would be identified for showcasing on the portal.

The scheme also includes free seats for existing students of higher education from Weaker sections of society. The distribution of free seats would be done through the NEAT portal, based on student information shared by Educational Institutions. During the first phase, the portal would be launched as a pilot phase in AICTE approved Government Colleges of India only.

5.14. Skill Assessment Matrix for Vocational Advancement of Youth (SAMVAY)

SAMVAY is a framework designed to assess and enhance the skills of young people for vocational careers, aligning them with industry needs and promoting employability. It aims to bridge the gap between the demand and supply of skilled manpower by assessing existing skills, identifying areas for improvement, and providing pathways for vocational advancement.

In order to make education relevant and to create 'industry fit' skilled workforce, the institutions recognized to offer skill-based courses have to be in constant dialogue with the industry and respective Sector Skill Council (s) to remain updated on the requirements of the workforce for the local economy. There shall be credit-based modular programmes, wherein banking of credits for skill and general education components are permitted to enable multiple exit and entry. This would enable the learner to seek employment after any level of Award and join back as and when feasible to upgrade her / his qualification / skill competency either to move higher in her / his job or in the higher educational system.

'Credit' is recognition that a learner has successfully completed a prior course of learning, corresponding to a qualification at a given level. For each such prior qualification, the student would have put in a volume of institutional or workplace learning, and the more complex a qualification, the greater the volume of learning that would have gone into it.

Credit framework is a pre-requisite to ensure mobility of learners with credit currency towards recognition of their acquired competencies. The credit framework provides for credit-based modular programmes, wherein banking of credits is permitted to enable multiple exit and entry. Specific credit-based assessment and award system has been incorporated giving sixty percent weightage to the skill component and forty percent weightage to general education component. The credit banking i,e. Credit Accumulation and Transfer System (CATS) allows a unit of assessment is a set of skill based learning outcomes, which can be of any size demonstrating learning outcomes for a NOS or Module.

A credit system is a framework with a set of specifications used by all institutions and certifying/awarding bodies within the national education and skill training system. The components identified are: Learning Outcomes, Assessment Criteria, Level of Learnings, and Quantum of Learnings. In the development of a comprehensive national system, a uniform and consistent definition of credit is required that will enable judgments to be made about the value of achievement, irrespective of time, place or mode of learning,

5.15. Skilling India at Global Standards

India's efforts in skilling at global standards are reflected through strategic initiatives such as the Skill India International Centres (SIIC) and partnerships facilitated through Government-to-Government (G2G) Memorandums of Understanding (MoUs). The establishment of 30 SIICs, as announced in the financial year 2024 interim budget, marks a significant step towards enhancing India's global skilling footprint. Currently operational centres in Varanasi and SDI, Bhubaneswar exemplify this initiative's early success, with plans finalized for seven more centres in the first phase.

India has also forged MoUs with prominent countries including Australia, Denmark, France, Germany, Japan, Qatar, UAE, and the UK for co-operation in Information Exchange, Standard Setting, Mutual Recognition of Qualifications, etc. Such partnerships not only enhance the international mobility but also promote the recognition of Indian qualifications abroad. Moreover, the establishment of NSDC International Limited in 2021, for ethical and transparent international recruitment of skilled Indians, drives the Skill India International Mission with a focus on priority sectors such as Information Technology, Construction, and Hospitality. The efforts include capacity building through 20 NSDC-affiliated training centres and language training at 12 centres, contributing to the deployment of over 26,000 skilled candidates across multiple countries.

5.16. Partnering with Industry for Skilling

Industry connection is crucial to any large-scale skilling programme, enabling contemporary relevance and employability and ascertaining demand to absorb the newly skilled workforce. In cognisance of this, the Skill India mission actively collaborates with the industry through National Skill Development Council driven partnerships for skill development, reskilling, and upskilling. Until March 2024, 131 projects have been undertaken by NSDC, with 62 corporate organisations benefitting over 3.10 lakh persons across the country, including 42 aspirational districts.

MSDE has signed 19 MoUs with its various departments, one led by the Directorate General of Training (DGT), 3 MoUs signed by the National Institute for Entrepreneurship and Small Business Development (NIESBUD), 5 under the Pradhan Mantri Kaushal Vikas Yojana (PMKVY), one under PM Vishwakarma and 9 MoUs through the National Skill Development Corporation (NSDC). These collaborations with the private sector and fellow ministerial departments play a pivotal role in transforming India into 'Kushal Bharat, Viksit Bharat' where the youth will not only be equipped to meet the demands of both, domestic and international markets but also will lead and innovate on a global scale.

DGT renewed its partnership with the Future Right Skills Network (FRSN), facilitated by Quest Alliance in collaboration with other partners Accenture, Cisco, JP Morgan, and SAP.

5.5 Skill Impact Bond

Skill Impact Bond is the first development impact bond in India dedicated to skills training and job placement. Launched in 2021, the **Skill Impact Bond** leverages an innovative and results-based finance mechanism to attract private sector funds and expertise for skill development, job placements, and retention. This initiative by NSDC and its coalition partners targets training 50,000 youth, ensuring at least 60 per cent are female, through selected and monitored NSDC-affiliated training partners over four years. Between November 2021 and March 2024, 29,365 candidates have been enrolled over five cohorts, 23,464 have been certified, 19,209 have been placed and 13,853 reported job retentions. The programme has reported 74 per cent women enrolment so far.

Source: https://nsdcindia.org/sib

References

Arora, M 2023The Strategy and Skill Value chain and its approach, https://www.linkedin.com>pulse>strategy-skill-value-chain

Eduquity Skill Assessment Services in India, http://www.eduquity.com>skill-assessment

MoLE National Policy on Skill Development, https://labour.gov.in>policies>national-policy-skill-development

MSDE 2025 Sector Skill Councils, https://www.msde.gov.in>organizations>ssc

Michelle Rea 2021 The Innovation strategy and Skill value chain approach, https://www.gatsby.org.uk>education>latest>the-innovation-strategy-and-skill

NIELIT About NSQF, https://www.nielit.gov.in>content>nsqf

NSDC Agricultural Skill Council of India, https://nsdcindia.org>nos-listing

NITI Aayog Skill development, Labour and Employment, https://www.niti.gov.in>verticals>skill-development-labour

NITI Aayog Fostering Skill Development and Learning Culture, https://nitiforstates.gov.in>best-practices-detail

Sanghi, M S and Srija, A 2015 Skill Development and the Productivity of the Workforce, http://www.niti.gov.in>sites>default>files

CHAPTER 6

SKILL DEVELOPMENT IN OTHER COUNTRIES- A GLOBAL PERSPECTIVE

Economic growth and the future of a country depends on the knowledge and skills of human resources. The skill force of a country requires advanced skill set training to be able to survive in the global era. In order to sustain in global competition, countries have to invest in education systems related to technical and vocational skills, which make their workforce more flexible and adaptable to meet the demand for a skilled workforce.

Though India has a demographic advantage compared to other countries, with respect to vocational education, India presents a gloomy picture. This has resulted in low labour force participation rate and high unemployment rate compared to some of the advanced countries across the globe. The students enrolled in skill development training and vocational programmes in India is only 1.4 percent when compared to 31.8 percent in the UK, 19.2 percent in Germany, 11.4 percent in Japan, 28.8 percent in Australia, and 18.9 percent in China.

6.1. Country-wise Skill Qualification Framework

The National Skill Qualification Framework is a key aspect in the skill ecosystem which defines the road map for skill development in the respective countries. The table below presents country wise skill qualification frameworks and its leading agencies with its implementing ministries from the selected countries in world.

Sl. No	Country	Qualification Framework	Lead Agency	Implementing Ministry
1.	India	National Skill Qualification Framework (NSQF)	National Skill Development Corporation (NSDC)	Ministry of Skill Development & Entrepreneurship, GoI

	Germany	German Qualification Framework (DQR)	German Office of Vocational & Educational Training (GOVET)	Federal Ministry of Education & Research (BMBF), Germany
	United Kingdom	Framework for Higher Education Qualification (FHEQ)	Quality Assurance Agency for Higher Education (QAA)	Ministry of State for Skills and Enterprise, UK
	Australia	Australian Qualification Framework	Qualification Framework Council, Australia	Department of Education, Skills and Employment
	China	Educational and Occupational Qualification System	Ministry of Human Resources & Social Security	Ministry of Education
	Japan	National Vocational Framework	National Institution for Academic degrees & University Evaluation (NIAD-UE)	Ministry of Education, Culture, Sports, Science and Technology (MEXT)

Source: Adapted from various reports

The skill development system and vocational education practices in the selected countries across the globe is presented below:

6.2. Germany

Germany largely follows a dual-system of vocational education and training (VET). The system is called "dual" because training, under this system, is conducted in two places of learning: in the enterprise and in the vocational school. The company provides practical training, and vocational school supplements this on-the-job-learning with theoretical instruction and basic economic background.

The dual system is especially well-developed in Germany, integrating work-based and school-based learning to prepare apprentices for a successful transition to full-time employment. The dual-system is the largest provider of education and training at the upper secondary level. Around 75 per cent of young people enrolled in VET take part in the dual system of vocational training.

The dual system has been a major factor in Germany's economic success and inventiveness over the past six decades. Many countries have looked at Germany's dual education system when reforming their own VET/ Skill education system.

The main characteristic of German VET is cooperation between companies (farms) and public vocational schools and it is complemented by inter-company courses.

Germany faces major challenges in terms of climate protection, quality of natural resources and biodiversity, which may vary regionally. In its CAP Plan, Germany therefore focuses on the mitigation of greenhouse gas emissions, the adaptation to the effects of climate change, the improvement of soil and water quality and on the enhancement of biodiversity.

Environmental Action Germany (DUH) influences the next generation of farmers regarding climate and health protection through trainings. Only an informed farmer, who knows about the impacts of his own actions and the right practical tools to avoid negative effects on environment and climate, will implement this knowledge in his own business.

The strengths of vocational education in Germany are:

- The industry receives low-paid trainees in a fixed time frame;
- Employers receive qualified personnel for the future generation;
- Students receive job security through on-the-job training on updated industry infrastructure;
- Vocational training in Germany is funded by the Government and trainees gets payment from the industry.

6.3.Australia

Australia's Vocational Education and Training is an integral part of the overall education system in the country designed to deliver workplace specific skills. The system has been very successful, reflected in high participation rate of adults in vocational training.

The Australian VET system is characterized by focus on lifelong learning and a modular structure. Education in Australia is compulsory for children between five and fifteen years of age. Anyone over fifteen years of age can access VET.

VET is a sophisticated system governed by interconnected government and independent bodies functioning within a National Skills Framework of Qualifications defined by Industry Training Packages and the Australian Quality Training Framework (AQTF).

In Australia, the approach to training and skill development is often defined as a market approach. Off the-job training was offered through state-owned and controlled "Tertiary And Further Education (TAFE)" institutes in the traditional system of "Vocational Education and Training (VET)".

The skills of agriculture and production horticulture industry workers are key to making Australia one of the most food secure countries in the world. Agriculture also plays a key role in the Australian economy, with around 70 per cent of primary produce exported. The industry is being shaped by numerous factors, including shifting global markets, changing consumer demand, labour shortages, COVID-19 disruptions, climate-related disasters, and escalating input costs.

Australia advocates a comprehensive approach to agriculture and food security that targets the immediate needs of the poorest, while also strengthening the foundations of agricultural industries through improving agricultural productivity, sustainability and opening markets. Australia recognises the role that climate resilient agriculture plays in addressing the complex problem of improving nutrition and encourages nutrition-sensitive agricultural investments.

The strengths of Australian VET system are:

- The Australian VET system empowers employees and employers to meet their training and skill needs by maintaining well-built relationships between Vocational Education Training and the labour market;
- Provision of transfer of credits for higher studies and other qualifications under national portability of qualifications and units of competency;
- Provision of Local autonomy and flexibility in the system
- Innovation to adapt learning and local conditions;
- Easy availability of information and research on vocational and education and training (VET) issues.

6.4.China

The Chinese Technical Vocational Education and Training (TVET) is a comprehensive system which is systematically planned to meet the needs of vocational education and training at different levels. To meet the challenges of China's industrialization drive, the government has repeatedly taken initiatives to reform China's TVET system. The Vocational Education Law of 1996 provides the legal framework for the implementation/functioning of the vocational education and training system in the country.

China's skill development model has been highly successful and developed at a rapid pace. In China, 59 per cent of those entering the workforce are skilled, of which 39 per cent are vocationally trained. China's skill development model has enabled transforming many of its high schools into skill-training centres, and successfully promoted internships for the vocationally-trained. Around one third of young people in China today enter vocational upper secondary schools.

In China, the expansion of education is closely related to its industrialization, economic growth and labour market demands. The government of China uses vocational training to promote equal opportunity for education through providing training for the unemployed, rural migrant workers, disabled people, farmers, and former soldiers.

Agricultural vocational education in China consists of secondary vocational education, higher vocational education and various formats of adult vocational education and training. Adult agricultural vocational education includes village-level classes to teach agricultural knowledge, production techniques and operational skills. The Central Agricultural Broadcasting School under the Ministry of Agriculture also provides training for farmers.

Chinese education policy is based on life-long learning and thus gives due recognition to adult literacy and training. For rural workers engaged in agriculture, government provides applied technical training in agriculture which includes green certification training and entrepreneurship training.

According to China's latest Catalogue of Encouraged Industries for Foreign Investment (2022 Version), foreign investors are encouraged to invest in multiple sectors related to the vocational training industry, such as:

- Non-academic vocational training institutions;
- Vocational schools;
- Elderly care service skills training;
- Family caregiving skills training; and
- Virtual Reality (VR) or Augmented Reality (AR)-based vocational and skills training services.

To capitalize on these prospects, foreign investors should carefully assess the Chinese market landscape, understand local regulatory dynamics, and establish strong partnerships with local stakeholders. By aligning their strategies with the China's vocational education goals and adapting to the specific requirements of the market, foreign investors can position themselves for success in the country's expanding vocational training industry.

The strengths of Chinese VET system are:

- Majority of the skill development programmes in China are highly decentralized run by the local governments. The local governments have freedom in deciding the curriculum based on the local needs.
- The industry participation in vocational education and training in China is ensured through the 1996 vocational educational law. According to this, each student should spend one year on workplace training during their upper secondary programme.
- The practical training at the organizations equips teachers with advanced tools and growing manufacturing requirements;
- Stipend is provided for vocational students to meet their needs, such as food and accommodation, making the tuition fee free of cost.

6.5. Japan

Japanese government introduced a farmer policy called new farmer support programmes, as below:

1. Agricultural employment programme for the corporation side (2008) - In this programme, government meets the training cost for new young generation interested in agriculture.
2. Young Farmer Benefits (2012) - In this programme, government provides assistance in farming preparation for independent and self-employed farming. The government provides loan for two year training and also gives financial support for ten years after starting farming.

Japanese agriculture faces aging problem and 50 per cent population of farmers are over 60 years. Hence, new technology and automation revolution is essential to Japanese agriculture. In present time, Japanese farmer use robotics technology, unmanned farming machinery, multipurpose tractor etc for farm works.

Japanese government is trying to attract more youth in the agriculture sector. They are taking several steps such as providing financial aid, subsidy, security, establishing agro-related industries, introducing new technology in agriculture, opening training centres related to agro works and promoting organic agriculture etc.

Skill development for foreign workers in Japan has largely been carried out through the "Technical Intern Training Program" (TITP). Originally intended to promote international cooperation in skill development, TITP is transformed into a system aimed at compensating for severe labour shortages.

Japan has proposed for the abolishment of TITP and suggested a new Skill Developing Worker Programme (SDW), accompanied by a comprehensive review of the entire system. The SDW is expected to offer a three-year employment period, allowing job changes after one or two years. Following the SDW, passing an examination would enable a transition to the Specified Skilled Worker Program (SSW) for approximately five years. Subsequently, transitioning to a higher-level SSW would provide not only job mobility, but also the opportunity to bring family members.

The strengths of Japanese VET system are:

- A broad-based universal core curriculum that creates enduring and supportive people with range of skills;
- Maintaining the educational system's inclusion through reducing dropout rates at each transition point;
- Increased connections between education and the labour market;
- Increased flexibility and individuality of youth by providing them with a larger range of choices.

6.6. United Kingdom

The Technical and Vocational Education and Training (TVET) sector in the United Kingdom is run by a number of different organisations, including colleges, companies, independent training providers, schools, and universities. Colleges are at the heart of the UK's TVET sector.

The United Kingdom is made up of four nations: England, Northern Ireland, Scotland and Wales. The UK Parliament is responsible for policy in England while the Scottish Government, Welsh Government and the Northern Ireland Assembly have a wide range of devolved powers, including over skills policy. In practice, they are similar TVET systems, but devolution has led to some interesting differences in how skills development is delivered and supported.

England: has introduced a training levy on large employers to fund increased apprenticeship numbers and is passing increasing responsibility to regions for determining skills needs to support economic development.

In England, Local Enterprise Partnerships (LEPs) have increasing influence over sub-regional skills policy, funding for adult skills development and capital expenditure. LEPs are led by employers who work in partnership with local government and other economic and social partners.

Northern Ireland: has also clustered and reduced the number of its colleges and has an integrated careers service model. It has pursued a policy of 'quality' modern apprenticeships. These schemes typically last three years and are focused on specific high level growth occupations.

Scotland: has maintained a stable TVET system with high quality apprenticeships and strong national careers information, advice and guidance services. It has clear skills plans for the major sectors of the economy, a consolidated college structure and a well-regarded credit and qualifications framework

Wales: has acted to reduce the number of its colleges. It has established regional skills partnerships, comprised of government, skill providers, and employers to address regional needs. Wales has also moved apprenticeship provision away from low level skills to focus on higher level qualifications. Government maintains a close direct relationship with key employers in Wales and works to bring them and training providers together.

To meet employers' needs, UK policy has a strong focus on engaging employers in both the design and delivery of TVET and ensuring the system is "demand led" rather than "supplier driven". This is helping to improve quality and relevance.

The strengths of VET in UK are:

- The skills system in UK is largely outcome based. The training providers have the flexibility to plan a delivery system based on the requirements of the trainee.

- The apprenticeship schemes in the UK form a vocational ladder that starts at 14 years age and ends with higher education level with employment;
- The UK skills system aims at making the dropouts ready for employment who is aged between 16 and 24 years.

6.7. Malaysia

Malaysia has made significant strides in modernizing its agro-food system. The agriculture sector in Malaysia needs higher productivity, efficiency, and resilience to support broader economic growth and more importantly, ensure Malaysia's food security amid both growing demand and increasing climate change impacts. Digital agriculture technologies (DAT) present a transformative opportunity for Malaysia's agro-food sector, promising enhancements in productivity, efficiency, and sustainability.

In order to become future ready, Malaysia launched Digital Skills for Tomorrow's Job programme, an initiative by Malaysia to prepare young adults across South East Asia for the 21st century workplace. This programme provides digital skills training for the most highly demand jobs in Data Analytics, Data Science, Development Operations, Software development and IT support. The programme assists Malaysia in upskilling and reskilling their digital skills for the most highly demand jobs.

Rapid Youth Success Entrepreneurship (RYSE) is an initiative launched by Malaysia, which empowers youth through entrepreneurship and employability programmes. It aims to train young people with transferable soft skills. Recognition of Prior Experiential Learning (RPEL) is an initiative with an aim to train working adults to upskill themselves in order to future-proof their jobs and become employable in the future of work.

6.8. USA

SkillsUSA is the largest workforce development organization for students in the United States. Representing over 413,000 career and technical education students and teachers, SkillsUSA chapters thrive in middle schools, high schools and college/post-secondary institutions nationwide. The mission is accomplished through the SkillsUSA Framework of Personal Skills, Workplace Skills and Technical Skills grounded in Academics, which is integrated into classroom curriculum.

Through the Framework, SkillsUSA students hone their hands-on skills against current industry standards in more than 130 occupational areas, from 3D Animation to Welding and nearly everything in between. At the same time, they develop the transformative career-readiness skills - teamwork, communication, professionalism, leadership and more- that fuel career and life success.

Skills Council of America has developed a skills ecosystem that supports an individual's journey to employment across all industries. The Skills Framework for Agriculture is becoming the globally accepted common language for the skills and competence for the agricultural world.

6.9. Singapore

Singapore's approach to skills development stands out globally for its comprehensive lifecycle perspective. Singapore has crafted an intricate ecosystem that supports citizens through three crucial career stages: school years, career entry, and professional growth. Singapore's approach to a skills-first human capital development system aligns closely with global initiatives such as the World Economic Forum's Reskilling Revolution. At the heart of this effort lies the Global Skills Taxonomy, a framework that establishes a common language of skills. This shared language is essential for aligning education, training, and labour market needs, ensuring that reskilling and upskilling efforts remain targeted and impactful.

The Institute of Technical Education (ITE) provides vocational education and training to students, equipping them with technical skills and knowledge needed in industries such as engineering, healthcare, and hospitality. Singapore's vocational training programmes emphasize industry relevance, practical training, and work-based learning. The country's commitment to vocational education has helped create a skilled workforce and attract foreign investment.

Launched in 2015, the SkillsFuture Movement addresses the various challenges such as an ageing workforce, rapid technological advancements, and the urgent need to transition to a digital and green economy.

The skills ecosystem starts with the demand side, which includes looking at what skills are likely to be required in the current and future workplaces. The Singapore Government has worked with tripartite partners to roll out the Industry Transformation Maps (ITMs) across 23 sectors to help companies and Singaporeans in these industries to identify future trends and manage industry disruptions.

SkillsFuture Singapore (SSG) has signed Memoranda of Understanding (MOUs) with five Trade Associations & Chambers (TACs) and Professional Bodies, to appoint them as "Skills Development Partners".

6.1 The Role of Reskilling Revolution

Launched at the World Economic Forum's 50th Annual Meeting in January 2020, the Reskilling Revolution aims to provide one billion people with better education, skills and economic opportunity by 2030.

The Reskilling Revolution aims to drive large-scale global change in employment and education by demonstrating how progress can be achieved on complex challenges. It focuses on piloting new models and templates for action, accelerating mechanisms for coordinating and funding collective efforts, and raising the standard for business leadership in times of major transformations. By reframing these agendas, the initiative seeks to inspire systems change and foster tangible, scalable solutions. With a focus on scalable solutions, it accelerates efforts in reskilling, upskilling, and workforce inclusion, setting a global example for replication and long-term impact.

Source: https://initiatives.weforum.org>reskilling-revolution

6.10. Korea

In Korea, agriculture is no longer seen as a career of last resort, but as an exciting opportunity for innovation, sustainability and personal fulfilment. The young farmers grants have sparked a trend of urban to rural migration, and are playing a crucial role in facilitating this transition, providing the necessary financial support and resources for urban migrants to establish themselves in their new rural environment. In the era of digital transformation, Korea's young farmers grants are also promoting the integration of technology in agriculture.

The Korea National University of Agriculture and Fisheries (KNUAF) has developed a unique programme that selects youth willing to take up farming and fisheries, provides them hands on experience with farming, support them in establishing a start up of their choice and provide continued support for another 7 to 10 years after they graduate.

KNUAF provides mechanisms both to create competitive manpower to manage high-tech farms and for continued handholding and financial assistance for trained graduates to become entrepreneurs. Field oriented and practical education is centred around practicality, business feasibility and field applicability.

6.11. Switzerland

Switzerland also has a highly regarded vocational training system known as the Swiss apprenticeship model. This model emphasizes practical training and collaboration between schools and businesses. It offers a wide range of apprenticeship programmes in various industries, allowing students to gain work experience while acquiring specialized skills. The Swiss vocational training system is closely aligned with the needs of the labour market and has played a significant role in Switzerland's economic success and low unemployment rate.

6.12. New Zealand

Agriculture Industry Training Organisation (AITO) was formed in New Zealand in 1997 to facilitate training for both new entrants and existing agricultural workers. The horticultural industry also formed the New Zealand Horticultural Industry Training Organisation.

The cadetship model for training in New Zealand enjoyed strong industry support with the Horticultural Cadet Scheme established in 1976, Equine scheme in 1982 and Pork scheme in 1985. These were collectively known as Primary

Industry Cadet Scheme (PICS). In New Zealand, 'marketing' has been replaced by an emphasis on quality assurance and supply chain management.

Agriculture in New Zealand has changed significantly over the past decade with less emphasis on sheep farming and strong growth in dairying and horticulture. The agricultural sector can be characterized by three main components- a) its prosperity is largely dependent upon international markets, b) productivity is comparatively high, fueled by technology and innovation, and c) many farmers in New Zealand are small, running family enterprises.

These cases highlight how different countries have implemented successful vocational training programmes tailored to their specific contexts and needs. While the specific models and approaches may vary, the common thread among these countries is a strong partnership between industry and educational institutions, a focus on practical skills development, and alignment with the needs of the labour market.

References

AOTS Skill Acquisition in Japanese enterprises,

https://www.aots.jp>publications>hrm.lr>report 03

CIPD Skill Development in the UK Workplace,

https://www.cipd.org>knowledgehub>Factsheet

Gangaiah and Srinivas, P 2018 Skill Development Practices: An International Context, Journal of Energy Technologies and Innovative Research, Vol 5, Issue 12, p 151-162

Jagannathan, S 2013 Skills Development: Promising Approaches in Developed Countries and Emerging Economies, ADB Briefs, No.19, Nov 2013

Kuruilla, S and Rodney Chua 2000 How do nations develop skills? Lessons from the Skill Development Experiences of Singapore, Cornell University, ICR School, https://core.ac.uk>download

MSDE 2024 MSDE partners with Australian Government to energize Agricultural Sector of India with emerging skills, https://pib.gov.in>PressReleaseframePage

NSDC Global Skill Gap Report: Assessing Overseas Opportunities for Skill manpower from India, https://sip.nsdcindia.org>files>kps-document

Ogden, A C., Streitwieser,B and Van Mol,C 2020 Education abroad: Bridging Scholarship and Practice, Routledge

Santhosh, M., Devi, K and Gandhi, A 2015 Understanding Skill Development and Training in China: Lessons from India, IAMR Report No 2/2014

SkillsFuture Singapore (SSG), https://www.skillsfuture.gov.sg

World Bank and IDA 2016 Skills Development in a Global context: Towards a future ready workforce, https://worldbank.org/en/news/speech/2016/08/15/skills-development

CHAPTER 7

SKILL DEVELOPMENT IN AGRICULTURE & ALLIED SECTORS

This chapter presents an overview of skill development trainings carried out by the various Ministries of the Government of India related to agriculture and allied sectors. It also attempts to provide a bird's eye view of the innovative initiatives undertaken by various state governments in the domain of skill development in agriculture.

7.1. Skill Development initiative by various Ministries

The table presents a synoptic view of the schemes of the various Ministries in relation to skill development in India.

Sl. No	Ministry	Scheme/initiative	Salient features
1.	Ministry of Human Resource Development	a. Apprenticeship under NAPS	- Provides support for practical training to 10+2 vocational pass outs based on CAC guidelines. Implemented through 4 Regional Boards
		b. Vocationalization of school education	-Provides for diversification of educational opportunities to enhance employability and reduce mismatch
		c. Community devt through polytechnics	- Employment oriented skill development programme, includes courses in agri and allied sectors

2.	Ministry of Agriculture & Farmers Welfare	a. STRY under SAME (Extension) b. Farm machinery training under SMAM c. DAESI d. KVKs e. Vocational training on empowering youth f. ARYA	- Implemented through SAMETI at State level & ATMA at district level, subsumed under ATMA cafetaria -Training on O & M of farm machineries in 4 FMTTIs -Diploma for agri input dealers organised by MANAGE - short duration technical trainings in agriculture and allied sectors - 21 days training by ICAR institutes - For gainful employment in selected agri enterprises
3.	Ministry of Micro, Small & Medium Enterprises	a. Entrepreneurship Skill Development Programme (ESDP) b.Multi-Disciplinary Training Centres	- To motivate youth to take up self-employment or entrepreneurship as one of career options
4.	Ministry of Rural Development & PR	a. DDU-GKY under NRLM b. RSETI	- Skilling & Placement initiative; Skill trainings imparted on various agri related enterprises - Run by banks with dedicated infrastructure, provide need based skill trainings
5.	Ministry of Tribal Affairs	-Vocational training for tribal youth in VTCs - TRIFED	- Each VTC run 5 trades imparting skill training to 100 or more tribal youth; to gain suitable employment or to become self employed -Capacity building for handicrafts/ handloom/ NTFP
6.	Ministry of Food Processing Industries	- Skill development in food sector - Food Industry Capacity & Skill Initiative (FICSI) - NIFTEM	-Provides sector specific skilled workforce from floor level workers to quality control supervisors in various segments of food processing - Offers skilling programmes leading to certification, placement, entrepreneurship focusing on food processing industry -Validation of QPs for identified job roles and curriculum development for the food processing sector

7.	Ministry of Social Justice & Empowerment	-CSS for development of SC living BPL -Special CA to SC Sub Plan	-Increase the income of the target population through various income generating schemes, skill development and infrastructure development.
8.	Ministry of Environment, Forest & Climate Change	-Green Skill Development -NCCF	Under ENVIS scheme, skill trainings are organised for rural youth in environment, forest and wild life
9.	Ministry of Fisheries, AH & Dairying	- Rashtriya Gokul Mission - NRLM -PMMSY	- Establishment of Multipurpose AI Technicians in Rural India (MAITRIs) - A-HELP (Accredited Agent for Health and Extension of Livestock Production) -Entrepreneurial development in fisheries like Ornamental fisheries
10.	Ministry of New and Renewable Energy	-Suryamitra under NISE -Varunmitra under NISE -Vayumitra under NIWE -Jal Urja Mitra under HRED	-Solar Power Technician - Solar Power Irrigation, water pumping - Wind Energy technician, wind pumps - Small hydro plant technician Skill trainings related to application of renewable energy in the farm sector
11.	Ministry of Electronics & IT	-Digital India programme -Future Skills PRIME	- Ensure digital access, digital inclusion, digital empowerment and bridging digital divide in the agricultural sector - Reskilling/upskilling of IT manpower for employability in emerging technology, including AI (covers agriculture also)
12.	Ministry of Women & Child Development	e-incremental learning approach Food & Nutrition Board	To tap the potential of technology to address the challenges in training Nutrition education and nutrition training; skill development in food & nutrition
13.	Ministry of Corporate Affairs	CSR funding related to skill development	

14.	Ministry of Personnel & Training	Skill Development through the i-GOT Platform	Mission Karmayogi- National Programme for Civil Services Capacity Building (NPCSCB) Capacity Building Commission (CBC) for functional supervision of training
15.	Ministry of Minority Affairs	Multi-sectoral development programme Seekho aur Kamao	-Improving socio-economic condition of minorities and providing basic amenities for improving quality of life -To reduce unemployment rate of minorities (Jammu & Ladak)
16.	Ministry of Labour & Employment	-Atmanirbhar Bharat Rozgar Yojana (ABRY) -National Career Service (NCS)	- Incentivize employers to create new employment opportunities - Career related services; NCS Portal, Model Career Centres, Interlinking of Employment Exchanges
17.	Ministry of Chemicals & Fertilizers	Institute of Pesticide Formulation Technology (IPFT)	Skill development for various stakeholders in chemical/agro-chemical sector
18.	Ministry of Jal Shakti	AIBP Har Khet Ko Pani (HKKP)	- Skill trainings for Repair, renovation and restoration (RRR) of water bodies
19.	Ministry of Science & Technology	DBT Nano mission	- Skill development for biotech enterprises - Nano fertilizers, nano pesticides, nano technology in food processing
20.	Ministry of External Affairs	India Centre for Migration	-Conduct Training of Trainers workshops and certify Master Trainers. Pre-Departure Orientation (PDO) training is imparted to migrant workers
21.	Ministry of Commerce & Industry	APEDA, MPEDA, GI registry	Skill development trainings to keep pace with changing customer preferences; marketing and promotion of 'Brand India'
22.	Ministry of Health & Family Welfare	FSSAI Food Industry Capacity & Skill Initiative (FICSI)	-Food Safety Training and Certification (FoSTaC) ; Resources like DART book offers simple tests for detecting food adulterants -Offers skilling programmes leading to certification, placement and entrepreneurship

23.	Ministry of Consumer Affairs	Indian Institute of Legal Metrology	Skill development in Legal Metrology
24.	Ministry of AYUSH	National Medicinal Plants Board	Capacity development on cultivation and protection of medicinal plants
25.	Ministry of Youth Affairs & Sports	Nehru Yuva Kendras	Skill development of rural youth (enhancing vocational skills) and facilitating access to skill trainings

Source: Compiled from various reports

7.2. Apprenticeship under NAPS

Apprenticeship is a crucial element in developing a skilled workforce in India. By offering on-the-job vocational training, it enables young individuals to acquire essential skills while earning a stipend. Globally recognized as a premier model for skill acquisition, apprenticeship programmes empower the youth with practical experience, enhancing their qualifications and readiness for the workforce. The Apprenticeship Programme in India is structured to include both Basic Training and On-the-Job Training (OJT) or Practical Training at the workplace.

To support and expand apprenticeship training, the Government of India launched the National Apprenticeship Promotion Scheme (NAPS). This scheme, managed by the Ministry of Skill Development and Entrepreneurship, provides financial incentives to establishments that undertake apprenticeship training.

Implementation of NAPS-2 in terms of candidate and establishment registration, creation of apprenticeship courses, posting of apprenticeship opportunities, contract generation and management, payment of stipend to apprentices through DBT and other activities related to the lifecycle of apprenticeship training will be online and through the apprenticeship portal (https://www.apprenticeshipindia.gov.in).

7.3. Vocational education in Agriculture

At the national level, out of 2,85,223 secondary and higher secondary schools (inclusive of Government, private aided and others), only 11,038 schools (3.86%) have vocational courses. (UDISE Plus - 2019-20 report).

The National Educational Policy of 2020 lays significant emphasis on the integration of vocational education in schools. Agriculture and allied subjects have a significant scope of being offered in the schools under the vocational stream. The students who have taken up agriculture and allied subjects including animal husbandry, fisheries, horticulture, sericulture, etc. as one of the subjects in their schools under the vocational stream will appreciate the prospects of the sector better and with appropriate promotional schemes, the chance of such students taking up agriculture as a subject for their higher education is expected to be high.

The Central Board of Secondary Education (CBSE) in India has included many vocational subjects in their senior secondary education. Poultry farming, horticulture and dairy science and technology are the agriculture based vocational subjects included.

The SAUs in Andhra Pradesh, Telangana, Himachal Pradesh, Kerala, Rajasthan, Tamil Nadu, Maharashtra and West Bengal are offering diploma courses in agriculture and allied sectors with varying duration (2-3 years) and eligibility (10th class / 10+2 level). Some private and public agencies are also offering certificate courses in agriculture for one year and above in local language. The employability of diploma holders is highly promising and are placed suitably.

7.1 Courses under VHSE

The following courses are offered in agriculture and allied sectors in the Vocational Higher Secondary Education (VHSE) schools in Kerala.

1. Agro Machinery and Power Engineering, 2. Agri-Crop Health Management, 3. Agriculture Science and Processing Technology, 4 - Agri-Business and Farm Services, 5- Livestock Management, 6 - Dairy Technology, 7 - Marine Fisheries & Seafood Processing, 8 – Aquaculture, 9. Floriculturist (Open cultivation), 10. Gardener, 11. Micro irrigation technician, 12. Dairy Farmer, 13. Floriculturist (Protected cultivation), 14. Agricultural Machinery Operator, 15. Small Poultry Farmer, 16. Dairy Processing Equipment Operator, 17. Fish and Seafood Processing Technician, 18. Ornamental Fish Technician, 19. Organic Grower, 20. Shrimp farmer.

Source: https://www.vhse.kerala.gov.in

Pandit Sunderlal Sharma Central Institute of Vocational Education (PSSCIVE) is an apex research and development organisation in the domain of vocational education in India. It is a constituent unit of the National Council of Educational Research and Training (NCERT) under the Ministry of HRD, Government of India. Agriculture and Animal husbandry form one of the six academic disciplines of the institute.

In order to fulfil the need of middle level skilled manpower in agriculture and allied sectors, the Department of Agriculture and Animal Husbandry (DAAH) of the Institute is involved in designing and developing vocational courses and teaching-learning materials in different core areas of agriculture and allied sector. Besides developing curricula, text books and practical manuals for plus two level vocational courses, the Department is also involved in modular competency based curricula and courseware for vocational subjects in schools under the NSQF.

The Institute is also a UNEVOC (International Project on Technical and Vocational Education) network centre in India, which co-ordinates all the activities, including communication with UNESCO-UNEVOC International Centre, Bonn,

Germany. Through the network, the Centre shares knowledge and experiences and discuss issues of common relevance in the vocational education field.

In addition to the skill development agenda undertaken by the MSDE, individual Ministries also undertake sector specific skilling and training initiatives (as evident in the table).There is considerable scope for aligning these different initiatives as well as for synergizing the implementation efforts. Unlike many other sectors, agriculture is not an aspirational job option either for farmers or people engaged in the sector in various capacities.

A brief description of the important schemes related to skill development in the farm sector is presented:

7.4. Skill development schemes for farmers

The Government is implementing Skill Training of Rural Youth (STRY) with the objective to impart short term skill training (7 days duration) to rural youths and farmers in agriculture and allied sectors for upgradation of their knowledge and skills and promote wage/self- employment in rural areas. The component aims at providing short duration skill based training programmes to rural youth and farmers on agri-based vocational areas for creating a pool of skilled manpower. Recently, the STRY programme has been subsumed under ATMA cafeteria.

The Government of India has identified about 50 skilling areas spanning Agriculture, Horticulture, Dairy science and Fisheries. The training activities are implemented through SAMETIs at State level and co-ordinated through ATMA at the District level. The training to rural youth is imparted through KVKs, Nehru Yuva Kendras (NYKs), Farmers Training Centres (FTCs) and such other identified training institutions at State/ District level.

Nehru Yuva Kendra Sangathan (NYKS) and Skill Development

Nehru Yuva Kendra Sangathan (NYKS) was set up as an autonomous organization under the Ministry of Youth Affairs and Sports, GoI to oversee the working of Nehru Yuva Kendras. NYKS is the largest grassroots-level youth organization, one of its kind in the world. It channelizes the power of youth on the principles of voluntarism, self-help and community participation.

NYKS has established a network of youth clubs in villages, where Nehru Yuva Kendras have been set up. NYKS has targeted to identify areas of harnessing youth power for development by forming Youth Clubs, which are village-level voluntary action groups of youth at the grassroots level to involve them in nation-building activities.

NYKS aims to upgrade the vocational skills of rural youth, enabling them to enhance their self-esteem and address issues in their daily lives. It assists youth in accessing skill development training and credit facilities under existing schemes.

The Government is implementing skill development programmes through Krishi Vigyan Kendra (KVK) under Indian Council of Agricultural Research

(ICAR) in different States of the country to serve as single window agricultural knowledge, resource and capacity development centres with mandate of technology assessment and demonstration for its use and capacity building.

A Centrally Sponsored Scheme on 'Support to State Extension Programmes for Extension Reforms' popularly known as Agriculture Technology Management Agency (ATMA) is implemented across the country by the Ministry of Agriculture & Farmers Welfare. The Ministry is implementing 'Sub Mission on Agricultural Mechanization' (SMAM). For implementation of this scheme, four Farm Machinery Training & Testing Institutes (FMTTIs) located at Budni (Madhya Pradesh), Hissar (Haryana), Geraldine (Andhra Pradesh) and Biswanath Chariali (Assam) are engaged in the country for imparting skill development training courses to different categories of beneficiaries like farmers, technicians, under graduate engineers, entrepreneurs on selection, operation, repair and maintenance, energy conservation and management of agricultural equipment.

MSDE, in collaboration with the Australian Government, hosted a comprehensive and productive round table to discuss the learnings from the Australia-India Critical Agriculture Skills Pilot Project. The findings of the scoping study were shared with MSDE and the Australian Government and subsequently 107 emerging job roles were mapped to identify the 5 critical emerging job roles in the agriculture sector for the pilot projects.

7.5. Skilling initiatives by various Ministries and organisations

The prominent skilling initiatives taken up by other ministries include:

- Ministry of Labour and Employment (MoLE)'s eSHRAM portal for creating a National Database of Unorganized Workers (NDUW) and eMigrate for the Overseas Employment Division of Ministry of External Affairs (MEA).
- Ministry of Rural Development: (MoRD)'s Deen Dayal Upadhyaya Grameen Kaushal Yojana (DDU-GKY) was launched in September 2014. The scheme focuses on diversifying family income of the rural poor and providing skilling and career support. The scheme emphasizes providing support to youth (15- 35 years old) in rural areas. Through this scheme, young people who have been unable to complete their education are able to obtain vocational training to build their careers. As of 2023, a total of 14.51 lakh candidates have been trained and 8.70 lakh candidates placed under the programme.
- National Urban Livelihoods Mission: NULM focuses on providing educational and skilling opportunities specifically to the urban poor. The main aim of this scheme is to reduce poverty and vulnerability of the urban poor by facilitating their access to different employment opportunities (including self- employment, wage employment, etc.), and access to shelter for those without homes and providing social security. According

to Annual report of NULM 2023, 1,92,122 candidates were skilled and trained, while 89,432 candidates were placed in jobs.

- Ministry of Textiles: Textile is the second largest employer after agriculture in the country. The ministry has an Integrated Skill Development Scheme for the Textiles and Apparel Sector including Jute and Handicraft. The scheme aims to fulfil the need for skilled manpower to the textile and related sectors. The scheme aims at targeting 15 lakh people over the span of five years. The Ministry of Textiles implement Integrated Skill Development Scheme (ISDS) to upgrade the skills of textile workers/ handloom weavers and Samarth (Scheme for Capacity Building in Textiles Sector).

As a part of the "Transformation of Aspirational Districts" programme, spearheaded by NITI Aayog and launched in 2018, there has been a special focus on skill training of youth in 112 Aspirational Districts, including those from vulnerable and marginalised sections of society, with various stakeholders and agencies coming forward to empower youth in aspirational districts.

To align qualification standards, the collaboration of Agricultural Skill Council of India (ASCI), and Skills Impact (Sector Skill Council equivalent industry body in Australia) is approved by the National Council for Vocational Education and Training (NCVET). As part of this project, there are six pilot projects nearing completion implemented in four states, Andhra Pradesh, Maharashtra, Tamil Nadu, and Telangana with new and emerging job roles in the sector that include Digital Agriculture Extension Promoter, Carbon Farming Practitioner, Livestock Green Management Promoter, Organic Farm & Business Promoter, and Integrated Farming Practitioner. The job roles are aligned with Sustainable Development Goals, primarily climate action, reduced inequalities, quality education and industry, innovation and infrastructure.

Australia has been a key partner for India and both countries are working together on capacity building and training, identifying opportunities in key sectors for closer skill cooperation with a particular focus on new-age courses as per the changing needs and demographics across both the Nations. Following NEP 2020, India has come up with new initiatives for promoting internationalization of education that include regulations for Joint/Dual/Twinning degrees and a draft regulation for setting up campuses of foreign universities in India.

Skill Development in Science and Technology

There are new age skills like artificial intelligence, Internet of Things (IoT), 3D printing, Virtual Reality (VR), fraud detection, smart sensor, cloud computing and Robotics process automation, which are gaining importance. These skills will equip youth to take up high-paying jobs in the domestic and international market. All the major skills development programmes in India like PMKVY, SHREYAS (Scheme for Higher Education Youth in Apprenticeship and Skills) cover this aspect.

Skill Development Training through Science & Technology (STST) is involved in developing special curricula and creation of models for innovative skilling areas.

Student Startup NIDHI award takes forward the student innovation in IEDC to commercialization and accelerate the journey of idea to prototype by providing initial financial assistance.

Augmenting Writing Skills for Articulating Research (AWSAR) disseminate Indian research stories among masses in an easy to understand and interesting format, it aims to empower and endow popular science writing skills among young PhD scholars.

Knowledge Involvement in Research Advancements through Nurturing (KIRAN) is an exclusive scheme for women, providing opportunities to women scientists who had a break in their career primarily due to family issue.

PM Vishwakarma Scheme was launched in 2023 to enable the recognition of artisans and craftspeople as Vishwakarma, making them eligible to avail benefits under the scheme. The scheme aims to support artisans and craftspeople who work with their hands and tools, including 18 trades such as carpenters, blacksmiths and potters. It has provision for skill upgradation to hone their skills, provide support for better and modern tools to enhance their capability, productivity, and quality of products and services, provide the beneficiaries an easy access to collateral free credit and reduce the cost of credit by providing interest subvention, provide incentives for digital transactions, provide a platform for brand promotion and market linkages to help them access new opportunities. The scheme aims at providing several benefits to the Vishwakarmas, who are either self-employed or intend to setup their own small-scale ventures.

The **RKVY-RAFTAAR** (Rashtriya Krishi Vikas Yojana- Remunerative Approaches for Agriculture and Allied Sector Rejuvenation) scheme is aimed at supporting agri-based startups and agricultural projects by providing them with the funding, infrastructure, and mentorship they need to scale.

The scheme primarily focuses on promoting startups and providing them with the tools to scale their operations. It encourages agri-entrepreneurs to develop innovative solutions for challenges in the agriculture and allied sectors. By supporting agri-entrepreneurs, startups, and organizations, this scheme helps integrate modern technologies, improve agricultural infrastructure, and create employment opportunities.

The scheme offers mentorship programs to help agri-entrepreneurs navigate challenges, optimize their operations, and scale their businesses. The participants also receive training programmes that focus on leadership, business management and technical skills.

The Skills Development Framework of the Department of Animal Husbandry and Dairying (DAHD) of the Central Ministry aims to facilitate, monitor and coordinate skills development activities within the country. The framework helps the department with a well-coordinated and coherent approach to skills development interventions. The framework provides a comprehensive approach to skills development within the livestock sector, catering to the diverse needs of stakeholders and contributing to the overall sustainability and growth of the sector in India.

The framework applies to a diverse range of stakeholders involved in the livestock sector, including but not limited to: Livestock farmers, Frontline workers, Entrepreneurs, Extension workers, Veterinary professionals, Veterinary paraprofessionals, Community Animal Health Workers (CAHWs).

The framework addresses skill development needs related to: Livestock farming, Animal breeding, Animal care and management, Dairy production, Animal health services, Outbreak monitoring, Disease Surveillance and Sampling.

National Livestock Mission intends to achieve the objectives of employment generation through entrepreneurship development in small ruminant, poultry and piggery sector and fodder sector, increase of animal productivity through breed improvement, increase in production of meat, egg, goat milk, wool and fodder. One of the objectives of the Mission is promoting skill based training and dissemination of technologies for reducing cost of production, and improving production of livestock sector.

The Department of Animal Husbandry, Dairying and Fisheries (DADF) of the Govt. of India has restructured the fisheries scheme by merging all the ongoing schemes under the umbrella of Blue Revolution. The restructured scheme provides focused development and management of fisheries, covering inland fisheries, aquaculture, marine fisheries including deep sea fishing, mariculture and all activities undertaken by the National Fisheries Development Board (NFDB), Hyderabad which is the executive arm of the ministry.

Under the Blue Revolution Scheme, NFDB has been mandated to undertake skill upgradation of stakeholders in the Fisheries sector, keeping in view the needs of the sector and broad funding patterns. NFDB was also appointed as Project Implementing Agency (PIA) to impart Recognition of Prior Learning (RPL) Training for Skill Development in Fisheries sector.

NFDB offers training programmes, awareness campaigns, and capacity building initiatives, particularly under the Pradhan Mantri Matsya Sampada Yojana (PMMSY). Skill Development Programmes (SKD) of NFDB focus on equipping individuals with specific skills relevant to the fisheries sector. Training of Trainers (ToTs) are organised by NFDB to train individuals who can then impart skills to others. Long Duration Skill Development Programmes of NFDB align with ASCI-QPs (Assessment Standards and Certification for Quality Programmes). NFDP aims to formalize the Indian fisheries and aquaculture sector by creating a database of work-based identities for all stakeholders. It undertakes CSR projects to address issues like unsafe fishing practices and lack of knowledge on life-saving equipment.

Indian Chamber of Food and Agriculture (ICFA)

While India has achieved self-sufficiency in food grain production, its agricultural productivity and quality standards often fall short of international benchmarks, leading to higher production costs and reduced global competitiveness. To address these issues, strategic global cooperation, innovation, and a comprehensive rethinking of policies and practices are vital to elevate Indian agriculture to meet global standards. It was against this background ICFA was created in 2015 with headquarters in New Delhi. ICFA views that agriculture has to be practiced on scientific lines using the best of technologies available globally, if it has to compete globally in cost and quality.

ICFA is the apex body in India, working on business, policy and development agenda and serving as global platform for trade facilitation, partnerships, technology and agribusiness services. With 27 industry Working Groups and sector specific Business Councils, ICFA represents the interests of key stakeholders at the national level and through its international platforms and partnerships, ICFA facilitates India's global engagement in food and agriculture sector. For enhanced sustainability and food safety and quality standards, ICFA has taken up the Agriculture Stewardship Programme by launching Healthy Food Initiative programme and Good Agriculture Practices (GAP) project.

National Dairy Development Board (NDDB) launched Operation Flood in 1970, which has helped dairy farmers direct their own development, placing control of the resources they create in their own hands. The bedrock of Operation Flood has been village milk producers' cooperatives, which procure milk and provide inputs and services, making modern management and technology available to members. Operation Flood was implemented in three phases. From the outset, Operation Flood was conceived and implemented as much more than a dairy programme. Rather, dairying was seen as an instrument of development, generating employment and regular incomes for millions of rural people.

NDDB offers training/orientation to milk producers, Board of Directors and procurement personnel of dairy co-operatives. Training is also imparted to the milk producers affiliated to New Generation Cooperatives. The training modules focus on the functioning of village-level dairy institutions as viable, vibrant and sustainable entities which provide reasonable and remunerative returns to milk producers. The emphasis is on good governance and professional management at all levels of the value chain of the milk business.

National Centre for Organic Farming is a nodal organization for promotion of organic farming under INM Division, Ministry of Agriculture & Farmers Welfare, Government of India. National Centre for Organic Farming has been renamed as **National Centre for Organic and Natural Farming (NCONF),** Ghaziabad in March 2022 with five Regional Centres for Organic and Natural Farming (RCONFs) located at Ghaziabad (North Zone), Bengaluru (South Zone), Bhubaneshwar (East Zone), Nagpur (West and Central) and Imphal (North East zone).

NCONF offers Residential Certificate Course on Organic farming. The programme is designed for Skill development of Rural Youth in field of Agriculture to develop entrepreneurs/Agriculture Startups in various sectors of organic farming. The programme is aimed to build capacity of rural youths linked to Local group/FPO/ICS members under PKVY/BPKP/MOVCDNER and any other State/central agencies scheme and strengthen organic food Industry especially in organic/natural production, quality base on-farm resource production and natural resource management and market value chain development.

Skill trainings by other institutions

National Horticulture Board (NHB) organizes various skill development programmes for farmers, entrepreneurs, and other stakeholders, focusing on production, post-harvest management, processing, and marketing of horticulture crops. NHB also focuses on developing entrepreneurial skills and leadership qualities in the horticulture sector. NHB has been designated as the Nodal Agency for the implementation of the Cluster Development Programme, which aims to promote integrated and market-led development of horticulture clusters.

Coconut Development Board organises skill development training programmes like Friends of Coconut Tree (FoCT) and handicraft training programme. FoCT training is one of the major skill development training programmes implemented in all coconut growing states through which massive employment generation is being created in coconut sector. CDB has developed Standard Operating Procedure (SOP) for skill training under FoCT. One week skill development training in Handicraft is also organized in different States through State Centres/Regional Offices.

Skill development for apiary in India focuses on training beekeepers in techniques like bee colony management, honey extraction, and disease control, with programmes offered by organizations like ICAR institutes and National Bee and Honey Mission (NBHM), aiming to enhance productivity and income for farmers.

To boost mushroom farming in India, skill development initiatives focus on training farmers and entrepreneurs in techniques like substrate preparation, inoculation, and harvesting, alongside value addition and marketing strategies, to enhance both production and profitability. Directorate of Mushroom Research, Solan, HP organises many skill development programmes in mushroom farming.

In India, skill development in sericulture focuses on empowering sericulturists and improving silk production through training programs covering various aspects like mulberry cultivation, silkworm rearing, disease management, and post-cocoon technology, aiming for sustainable and higher income. The Central Silk Board (CSB0 plays a lead role in organizing training programmes at the Central Sericultural Research and Training Institute (CSRTI) in Mysore, catering to various stakeholders in the silk industry.

The Ministry of Tourism implements a skill training programme for tourism service providers under the Capacity Building for Service Providers (CBSP) Scheme to enhance the employability of the candidates. This training includes fresh candidates, candidates for re-skilling and upskilling who are already working in the hospitality sector. This training programme is on pan-India level including the tier-II cities.

The Ministry of Tourism had initiated the Destination Based Skill Development training programme w.e.f. FY 2019-20, with the objective to bring in an all-encompassing training plan at the tourist destination itself to upgrade and build capacities of all service providers to the tourists.

The Institutes of Hotel Management (IHM) and Food Craft Institutes also organise skill based trainings. The following skilling, re-skilling and up-skilling programmes are conducted under CBSP Scheme:

i. Hunar Se Rozgar Tak (HSRT) (Skilling Programme): 'Hunar Se Rozgar Tak' Programmes under CBSP Scheme are compliant with the Common Norms notified by MSDE, Government of India for various Ministries for implementing skilling programmes. The programme presently offers a total of eleven short duration courses in hospitality and tourism sector.

ii. Entrepreneurship Programme (Up-skilling): This programme is to facilitate micro and small business start-ups.

iii. Skill Testing & Certification (Re-skilling): This programme is for the existing service providers to test and certify them in four hospitality trades.

iv. Tourism Adventure Courses (Re-skilling): This programme is to develop capacity in youth to promote regional based specially structured adventure skill.

v. Linguistic Tourist Facilitators (Up-skilling): To create trained manpower in different foreign languages.

vi. Tourism Awareness /Sensitization programme: This programme is conducted for existing service providers.

vii. Destination Based Skill Development: This programme is implemented with the objective to sensitize, upgrade and build capacities of all tourist service providers, especially to the people who are residing near tourism sites and destinations and to enhance employability.

7.6. Skill development initiatives by State Governments

The innovative skill development projects related to agriculture and rural development undertaken by the some of the state governments is presented below:

Maharashtra

Maharashtra State Skill Development Society (MSSDS) is the single Nodal Agency for planning, coordination, execution and monitoring of skill development initiative of Government of Maharashtra. A state run programme titled Pramod Mahajan Kaushalya and Uddyokakta (PMKUVA) is aimed to increase employment awareness and skill development in the state. Under the state's flagship programme PMKUVA, skill training is provided to adolescents in a variety of vocational and technical courses through Industrial Training Institutes.

In October 2018, the NSDC and the MSSDS launched an ambitious agri-skilling programme, under PMKVY 2.0. **The Maharashtra Agri-Skilling Programme (MASP)** was one of the first Special Projects implemented in India under Recognition of Prior Learning scheme to be implemented at massive scale and speed, with the aim of training and certifying 281,000 farmers across rural Maharashtra. This innovative intervention by the Maharashtra government goes a step beyond certifying the farmers, and facilitates farmers to express their interest to form Farmer Producer Organisations (FPO) while also offering them guidance for company registration, Business Advisory Services (BAS) for business planning, support in submission of their business plans to concerned financial institutions for credit/capital building. The MASP also comprises an individual skilling component, with its focus on enhancing skills and employability of youth in agriculture sector to drive the FPOs established by participants of the mass skilling intervention.

The programme was funded by the NSDC and implemented by the MSSDS in the state as the Project Implementation Agency (PIA). The Agriculture Skill Council of India (ASCI) developed the curriculum and assessment framework for the programme and the Maharashtra State Council for Vocational Training (MSCVT) was entrusted with farmer assessment and certification.

The MSSDS appointed a consortium of SIMACES Learning LLP and Palladium Consulting India Pvt. Ltd. as the official Training Partners to design and deliver farmer training and facilitate undertakings for FPOs, combining the synergies of both public and private sector to achieve optimum positive impact for farmers in Maharashtra.

Under Chhatrapati Rajaram Maharaj Entrepreneurship and Skill Development Mission Special Agri Project (MSDP), the Government of Maharashtra has identified agriculture as a crucial sector to focus on enhancing Maharashtra's future productivity and economic competitiveness. The Government identified that triggering new employment and developing market opportunities across agricultural value chain is critical.

The Skill Development and Entrepreneurship Department of Maharashtra in collaboration with UNDP has pioneered a unique initiative called 'Skill Sakhi' that focuses on the most critical aspect of India's skill ecosystem, i.e., mobilization, handholding and mentoring. The initiative promotes mobilization and skilling of employable girls and women in the peri-urban and rural areas through network of young women community leaders.

Meghalaya

The Government of Meghalaya is engaged with the Asian Development Bank (ADB) in skill development through a dedicated project titled "Supporting Human Capital Development in Meghalaya". The project makes efforts and allocates resources to address the growing priority of youth employment and the link between skills training and jobs.

The British Council team works with Meghalaya State Skill Development Society (MSSDS), focusing on capacity building of selected Programme Implementing Agencies (PIAs) working with MSSDS. The PIAs will be nurtured to grow and as the project moves ahead, ways will be explored to encourage them for tie-ups with other smaller training providers and to increase their base, reach, capacity and efficiency. Skills Challenge Fund (SCF) is created, which denotes a financing mechanism whereby funds are made available to eligible institutions or service providers to undertake specified activities aimed at overcoming or confronting a defined challenge.

The "Basic Agriculture Training Centre Scheme" is a State Plan Scheme implemented by the Department of Agriculture & Farmers' Welfare, Government of Meghalaya with the objective to provide short-term vocational training for rural educated unemployed youths in agriculture and allied sectors to develop entrepreneurship skills and employment generation.

The State Council of Science and Technology and Environment (SCSTE) in partnering with Bio-Resource Development Centre (BRDC) implement the Skill Vigyan Programme sponsored by the Department of Biotechnology, Government of India for the benefit of young graduates in Meghalaya. The objective of the Skill Vigyan Program is to impart hands on training using tools and technique in advance and emerging areas of life sciences and biotechnology and developing entrepreneurial skills to 10+2 students and graduates ready to take up jobs as well as to established start-ups, incubators on their own in their respective areas. The skill courses under this category were designed in consultation with experts of Sector Skill Councils and approved by MSDE.

Andhra Pradesh

Andhra Pradesh has structured an entity (APSSDC) on a PPP mode along with NSDC and private partners to initiate and oversee skill development efforts in the state. Skill trainings on Employable Skill Enhancement Module have been initiated aiming at beneficiaries of Mukhya Mantri Yuvanestham, a flagship programme of Andhra Pradesh providing unemployed youth allowance.

APSSDC has organised skill building trainings across a range of disciplines and also made placements in conjunction with regional firms and foreign investors. Under APSSDC, the Government of Andhra Pradesh has formed seven different Missions to achieve double digit growth for the state and to make AP among the most developed state in the country. Among these, the Knowledge and Skills Mission has been formed to provide trained and skilled manpower to all other Missions.

Society for Employment Generation and Enterprise Development in Andhra Pradesh (SEEDAP) is one of the largest skilling organizations in the country and the Nodal Agency for implementing the various Central and State Government's skilling programmes to facilitate the settlement of youth in Wage Employment, Self-Employment and Entrepreneurship. SEEDAP works with an objective to provide at least one employment opportunity in the organized sector to every poor rural household by conducting employment oriented, market driven skilling programmes to rural unemployed youth aged between 18 to 35 years focusing predominantly on school and college dropouts. Andhra Yuva Shakti (AYS) is implemented by SEEDAP to nurture the young talent to become entrepreneurs and self-employed in agriculture and allied activities.

In Andhra Pradesh, Rythu Sadhikara Samstha (RySS) was established as a not-for-profit company to create integrated institutional mechanism for all programmes, schemes and activities intended for farmer's empowerment, encompassing welfare, development, capacity enhancement, credit flow, financial support and allied empowerment activities. Andhra Pradesh Community Managed Natural Farming (APCNF) is being implemented by RySS. APCNF has been recognized for its significant contributions to global food security, climate resilience and ecosystem protection. This initiative specifically supports smallholder farmers with a focus on empowering predominantly women farmers.

Andhra Pradesh is pioneering a first-of-its-kind initiative in India with the launch of its Skill Census, aimed at enhancing the state's workforce capabilities. Central to this initiative is the NAIPUNYAM portal, an innovative end-to-end digital solution designed to streamline citizen services through three key features: assess, learn, and apply. This comprehensive approach not only facilitates skill assessment and training but also enables individuals to effectively apply their skills in the job market.

Telangana

Employment Generation and Marketing Mission (EGMM) is a state level society formed under the Rural Development Department to co-ordinate the efforts of employment initiatives in rural areas in Telangana. EGMM has emerged as one of the largest Jobs mission working for the underprivileged youth. As on May 2024, it has trained 2,26,909 and placed 75 per cent of them in entry level corporate jobs. 45 per cent are girls and 37 per cent are SCs/ STs. The youth are from economically and socially underprivileged sections of the society. EGMM works in a public-private partnership mode with Government, companies and the rural communities as its stakeholders. The entire approach is a bottom-up approach, tailored to move the rural poor from the unorganized to the organized labour market.

Telangana Academy of Skill and Knowledge (TASK) is established with the mandate of skilling the youth of the state and making them more employable. TASK is continuously evolving and bringing new programs with innovative pedagogies under its umbrella to address all the requirements of the 21st century workplace.

Telangana Scheduled Castes Co-operative Development Corporation Ltd (TSCCODC) was established to take up programmes for the economic development of Scheduled Caste families in the State. The Corporation provides financial assistance for creation of income generating assets and offers training programmes for Skill upgradation leading to self -wage employment.

Dr. D. Ramanaidu Vignana Jyothi Institute of Rural Development is focusing upon six months agri based vocational training in Medak district. The institute has tie-ups with several private companies like Nuziveedu seeds and Beyar which provides placements for students. The institute focuses upon the rural youth and small scale industries and seeks to address rural urban disparities, which are a serious concern to governance.

Bihar: Kaushal Yuva Programme (KYP)

It was observed that the youth of Bihar though qualified, lacked the basic soft skills and Information Technology (IT) skills which are a pre-requisite for any job in today's world. Because of this, the youth found it difficult to find the right jobs that match their qualification and aptitude. In this context, Bihar Skill Development Mission (BSDM) conceptualized and launched the Kushal Yuva Programme (KYP), a unique skill training programme to enhance the employability of the youth of Bihar which would add further value to the various domain specific training endeavours of the State.

The KYP in its initial phase targets the youth in the age group of 15 to 25 years who are at least 10th pass. The course curriculum is covered in 240 hours and includes three components, specifically designed to enhance employability viz. Soft Skills for 40 hours, Communication Skills for 80 hours and Basic computer literacy for 120 hours.

The mode of delivery is completely through an 'E-Learning' platform named as 'E- learning Revolution for All (ERA)', which features self-paced learning, interactive sessions, practical learning, and corrective mechanisms, continuous and final evaluation. The mode of learning is sequential and integrated into an online assessment and certification process, ensures that a candidate receives session-wise certificates after successfully completing each session and a final certificate on successfully clearing the final assessment.

To ensure that the benefits reach the doorstep of each and every needy youth, BSDM has constructed 534 Block Skill Development Centres (BSDCs) at each of the 534 blocks across 38 districts of the state. To run these BSDCs, BSDM has successfully empanelled reputed training organisations through a Request for Proposal (RFP) mode. Apart from BSDCs, BSDM has also registered private skilling centres with adequate space for running the KYP programme. An online registration process ensures uniformity and robustness across the centres.

Further, to maintain uniformity in delivery of training / pedagogy, BSDM has engaged in conduct of online Training of Trainers (TOT) and an online examination, ONCeT. All Learning Facilitators / Trainers have to mandatorily pass the ONCeT examination without which they are not allowed to conduct the training programme.

Chhattisgarh: Livelihood Colleges

The Chhattisgarh Right of Youth to Skill Development Act, 2013 ensures every person in the age group of 14 to 45 the right to develop his/her skills from among notified skills. District Skill Development Authorities set up under District Collectors are bound to provide skill development training within 90 days of receiving any demand in this regard.

Government of Chhattisgarh facilitates large population of unemployed youth/tribal population living in remote and under-served areas in States to undergo livelihood training to increase their employability / ability to earn a livelihood by providing residential training facilities through Livelihood Colleges.

7.2 Livelihood Colleges in Chhattisgarh

The Chhattisgarh Government in July 2013 approved "State Project of Livelihood Colleges" for free skill development training to the youth. The programme involves skill development facilities under one roof.

Training is provided in short-term, market -oriented, demand driven MES Courses. It facilitates multi-entry, multi-exit, multi-skilling with linkages to future up gradation opportunities. There is assistance for post training placement and very less entry barriers.

With the objective of placing the Livelihood College initiative on a robust foundation with strong employment linkage, and to create a network of such Livelihood Colleges to provide a range of livelihood opportunities for students from across the state including in partnership with the private sector, the State Government established the State Project 'Livelihood College Society'.

The training costs are met from the State's skill development Convergence Scheme, Mukhya Mantri Kaushal Vikas Yojana, in which skill development schemes and funds of various departments are being converged. In addition, since 2015-16, budgetary support is also given for skill development training.

Different models are being taken up by the Livelihood College Society for achieving the goal of employment-linked skill development for the youth of the State viz. departmental training, training through reputed private sector partners, and training under CSR by established players. In involving private players, the society is offering physical infrastructure, mobilization and facilitation in return for a commitment towards employment from prospective partners.

So far, 27 Livelihood Colleges have been established in 27 Districts to provide quality training, infrastructure and ensure presence of reputed training partners in each district. Skill Training is provided absolutely free of cost to the beneficiaries.

Source: https://www.nitiforstates.gov.in>best-practices-detail

Kerala

Kerala has taken several measures to encourage skill development in the state. The Additional Skill Acquisition Programme (ASAP) was introduced for those studying in various colleges in the State. ASAP strives to transform the State of Kerala to a human resource hub of the region, which is well positioned to capture growth opportunities. ASAP aims to impart industry-specific skills to the needy students in the selected schools and colleges along with their regular studies through modular courses with the active participation of industry partners. Students enrolled with ASAP get opportunity to learn sector specific skills of their choice. Communicative English and Basic IT modules are integral component of ASAP.

Skill Updating Institute for Industrial Training (SUIIT)

SUIIT was started to update the technical knowledge of Instructional Staff and others in pace with technical developments. The activities of this institute are preparation of course design, research in new methods of teaching, skill updating programmes of technical staff, soft skill training, training programmes under State training policy at IMG, off campus training programmes, and monitoring of virtual class training.

7.3 Kerala Academy For Skills Excellence (KASE)

KASE is a Section 25 company fully owned by Government of Kerala. The objective of this company is to facilitate mass human resource development through specialized skill development courses across the state. The participation of leading public and private industries is envisaged in the domains of construction, retail, finance, IT & ITES, tourism, and hospitality. The main purpose of the academy is to spearhead the creation of highly skilled and safe professional workers, for all the levels of work in the industry, through scientifically designed specialized courses for various categories of workers. Training will be imparted to sharpen the technical skills as well as soft skills. To be at par with international standards, the workers are trained in soft skills like communication skills, language skills, safety and environment awareness , which is currently lacking in our work force as well as professional course students, who graduate from various colleges. It aims at imparting theoretical as well practical training of skill sets needed for each category -from supervisory level to working class.

Source: https://www.kase.in

The District Skill Development Centres (DSDCs), earlier known as **'Kaushal Kendras'** is a scheme designed and implemented by the Department of Labour and Skills in Kerala with the intention of creating opportunities for skill development, for various sections of people from different backgrounds, including those from rural backgrounds. DSDCs are permanent systems established for better planning of skill training, taking into account the needs of the employers and the preferences of the job seekers to address the local labour challenges at the district level. DSDCs ensure skill training at local levels in various domains and also act as district-level facilitation centres for providing district-based skill information.

Tamil Nadu

Tamil Nadu Skill Development Corporation (TNSDC) was formed with a vision to transform the state into skill hub by skilling the youth to enhance their employability and match the expectations of the industry. Since 2013 it is functioning as a Corporation under the Special Programme Implementation Department. TNSDC has been declared as the State Nodal Agency for Skill Development by the Government of Tamil Nadu.

For improving the economy of the unemployed rural youth and for repair and maintenance of the agricultural machinery by the farmers in their own fields, Agricultural Engineering Department along with the Tamil Nadu Skill Development Centre is providing training in the workshops of Agricultural Engineering Department at Coimbatore, Vellore, Tiruvarur, Trichy, Madurai and Tirunelveli to unemployed rural youth in repair and maintenance of agricultural machinery and solar power pump sets.

TNSDC facilitates free short-term skill training programmes across a wide range of industries to empower youth with employment-oriented skills. These initiatives are delivered through more than 350+ approved training centres, covering over 38 sectors. Green jobs is one of the sectors covered under this.

Karnataka

The Department of Skill Development, Entrepreneurship and Livelihood was created in 2016 to bring all skill development activities under one umbrella to promote the development of the youth of the State, better growth of working hands and inclusion of socially inclusive groups.

Karnataka Vocational Training and Skills Development Corporation (KVTSDC) was set up for furthering the agenda of skill development in Karnataka.

The Government of Karnataka has constituted the Karnataka Skill Development Authority (KSDA) under the Department of Skill Development, Entrepreneurship and Livelihood in order to give a centralized role in the matter of standardization and quality skills training. KSDA is responsible for the establishment of quality standards in skills training. The Authority prepares the curriculum, course content, and credit frameworks for skill training. It lays down guidelines for training of trainers, prepares standards for skill training, and also improves standards of the skill programs in the State.

The Authority also designs and develops labour market information systems based on global standards and best practices. It works as a think-tank forum for research inputs relating to skill development. The Authority strives hard to fulfil the growing need in the State for skilled manpower across sectors and narrow the existing gap between the demand and supply of skills.

The Chief Minister's Kaushalya Karnataka Yojane (CMKKY) is the primary initiative in Karnataka aimed at providing short-term vocational skilling for youth, fostering employment, self-employment, and entrepreneurship. Open to individuals aged 18-35, the scheme offers free skill training in approximately 3,000 NSQF-aligned job roles across 35 sectors, including apparel, automotive, BFSI, IT & ITeS, healthcare, electronics, agriculture, retail, and tourism. With over 800 accredited CMKKY centres state-wide, the training adheres to industry standards.

7.4 Skill on Wheels in Karnataka

Targeting 0.5 million rural youth at taluka level in Karnataka, NSDC had launched Skill on Wheels campaign in 2017 in association with Raman International Institute of Information Technology (RiiiT), India's first IT finishing school (ITFS) and supported by GSS and Dharti Foundations. Some of the activities covered were Skillathon (to create skill awareness); Techathon (hackathon for programming and computer science enthusiasts); Teachathon (concept focusing on teachers to bring in innovation by blending technology and knowledge); 3I Summit (Industry, Institute and Investor summit bringing everyone on one common platform); Ideapreneur (entrepreneurship development program) and RISE (Rural Inclusivity to Strengthen Employment).

The Skill on Wheels covered 30 districts, 129 talukas in Karnataka reaching out to over 0.25 million youth and job seekers. Skill on Wheels is an impactful and a scalable concept as it reaches out to rural India effectively. It aims to bridge the gap between mentors and last mile candidates by spreading awareness about importance of skill development and providing success stories, all packaged in a skilling centre on wheels (state-of-the-art vehicle)

Source: https://www.deccanherald.com>India>Karnataka

Tripura

Directorate of Skill Development (DSD) was set up under the Department of Industries & Commerce, Government of Tripura, to bring more focus and coordination to the skill development initiatives under Tripura Skill Development Mission.

The Tripura government is actively fostering skill development through programmes aimed at youth empowerment in sectors such as agriculture, handicrafts, and tourism. The Tripura Skill Development Mission, along with initiatives like DDU-GKY, plays a key role in improving employability in rural areas. Private sector contributions, particularly from NGOs like Sankalp, further support these efforts by offering vocational training to marginalized communities.

Recognizing that skill up-gradation alone may have limited impact on the livelihoods of SHG members in Tripura, Livelihood and Enterprise Development Programme (LEDP) was launched, which focuses on creating sustainable livelihoods within SHG communities through a holistic approach. The LEDP scheme is designed to conduct livelihood promotion programs in clusters, incorporating various elements to ensure optimum benefits from skill up-gradation. The programme includes provisions for intensive training, refresher training, establishing backward-forward linkages, and providing handholding and escort support. Covering the entire value chain, LEDP offers an end-to-end solution to SHG members.

Gujarat

As a prominent participant in national skilling initiatives, Gujarat has established a strong network of training centres under the PMKVY and state schemes under the Gujarat Skill Development Mission (GSDM) - Mukhyamantri Bhavishya Lakshi Kaushal Vikas Yojana (MBKVY) and Saksham. Kaushalya Vardhan Kendras (KVKs) were established with a goal of providing employment to the youth of Gujarat. It imparts skill trainings to youth through village cluster training centres in the villages.

Kaushalya Vardhan Kendra is an initiative which offers a wide range of skill development training programmes across various sectors, including healthcare, tourism, agriculture, construction, manufacturing, and service industries. The initiative aims to bridge the gap between skill development and employment by facilitating job placement for trained individuals. It collaborates with industries, businesses, and organizations to create employment opportunities and connect trained individuals with potential employers. The programme specifically focuses on empowering women and aims to enhance their economic independence and promote gender equality.

Local Institute for Vocational Education (LIVE) provide localized training on sectors like IT, hospitality, Drone technology, Media and Entertainment, Manufacturing, Service and Banking & Finance etc. Private partnerships, industries partners and government training providers are pivotal in enhancing vocational training quality and addressing industry-specific needs.

West Bengal

The Technical Education, Training & Skill Development (TET&SD) Department is the nodal department for all the skill development interventions of the Government of West Bengal. All the skilling interventions of other departments are channelized through the TET&SD department.

Utkarsh Bangla scheme was launched for the entire Skill Development interventions to provide wage/self-employment linked skills training to the residents of West Bengal. It is implemented in public-private partnership mode with Private Training Providers (PTPs) and PTP-ITIs empanelled with Paschim Banga Society for Skill Development (PBSSD). Currently there are more than 500 PTPs covering trainings in more than 2400 Training Centres across 23 districts of the State.

Paschim Banga Society for Skill Development (PBSSD), an autonomous society under the TET & SD Department, has been formed with an aim to implement non-institutional skill development interventions in the state and ensure uniformity in the Training curricula / content, delivery, duration, Assessment & Certification procedures and Training cost.

The Skill Development Programmes under PBSSD can primarily be categorized in three categories: a) Entirely State Funded: Project Mode, Recognition of Prior Learning under Utkarsh Bangla b) State -Central Sharing: Deen Dayal Upadhyaya Grameen Kaushalya Yojana c) Entirely Centrally Funded: PMKVY-CSSM, Sagarmala, etc.

Rajasthan

The Raj-Kaushal portal established by the Government of Rajasthan has a vast database of skilled, semi-skilled and unskilled workers and also contain information on various employers registered as establishment, factory, industry, company, etc in the state. Envisaged as a step towards Aatmanirbhar Bharat, while the portal provides citizens opportunities to upgrade their skills and start working in nearby localities, it enables employers to search for qualified workers as per the required skill set. The portal serves as an integrated platform, wherein an employer can post the requirement and a job seeker can search various job opportunities.

Rajasthan's skill development policy is a forward-looking strategy that seeks to build a skilled, adaptable, and resilient workforce. As the state continues to implement and refine its skill development initiatives, it contributes significantly to not only individual empowerment but also to the broader goals of economic prosperity and social equity within the state. The policy covers many new initiatives and programmes of the Rajasthan government towards skilling its population.

Uttar Pradesh

The Uttar Pradesh Skill Development Mission (UPSDM) was launched to enhance employability of the aspiring youth through skill training. Six schemes of six ministries converged with the State Skill Development Fund to initiate the programme.

- Training was provided under Central scheme 'One District, One Product' that aims at encouraging local products.

- Training partners were roped in with a mandatory commitment to ensure 60% assured placement.
- Overseas employment was facilitated through the Non-Resident Indian (NRI) department and tie-ups with leading placement agencies and job portals.
- For training and placement of candidates, flexi memorandums of understanding (Flexi MoUs) were signed with 19 leading industrial entities such as Raymond, Maruti India, and Larsen & Toubro.
- A State Skill Development Fund (top-up fund) was constituted for meeting additional expenses born in integrating various initiatives of skill development.

The platform 'Sewa Mitra' was launched by the Employment Directorate under the Department of Labour, Uttar Pradesh in March, 2020. The platform employs a multi-dimensional approach leveraging technology as well as brick-and mortar institutions to create better reach. Sewa Mitra acts as a bridge between citizens and local service professionals required for day-to-day work or home services based on geo location in both urban and rural areas. Since its launch, Sewa Mitra has been a hit. Several skilled workers were provided employment by various industries. Many unskilled workers got employment under MGNREGA and many were provided skill trainings. After skilling and certification, workers enjoy a premium on the platform .

Odisha

Odisha Skill Development Authority (OSDA) acts as an apex body for monitoring , co-ordination, convergence and providing overall policy direction for skill development activities in Odisha. Training of Trainers (ToT) programme with technical support of Australian Skill Development and Commonwealth Educational Media Centre for Asia was initiated by the Odisha Government.

India's first World Skill Centre (WSC) was established in Bhubaneswar in March, 2021. WSC is a major milestone in establishing 'Skilled-in-Odisha' as a global brand. The Centre is being set up with a loan from ADB and technical assistance from ITEES, Singapore. WSC will house a Skill Museum and an advanced facility for training of trainers.

Uttarakhand

In Uttarakhand, the Integrated Livelihood Support Program (ILSP) has established livelihood collectives (LCs) and producer groups (PGs) to enhance farmers' income and employability. There are over 230 LCs and more than 13,500 PGs in the hill districts. A joint initiative, "Together We Can," involved LCs and PGs acting as nodal agencies for delivering groceries to villagers' doorsteps, ensuring fair prices and reducing market shortages. It also promoted the "vocal for local" concept, encouraging members to sell fresh produce in their communities, minimizing transmission risks and travel expenses.

Madhya Pradesh

Madhya Pradesh conducted a survey of migrant labourers who had returned to the state after Covid-19. Skill mapping was carried out in over 100 trades during the survey. To provide employment opportunities to these returned migrants, the state government launched the Rojgar Setu portal. The portal allows employers and employment suppliers to register and search for eligible employees among the migrant worker data uploaded by district administrations. The portal also monitors the provision of ration and welfare benefits to the families of migrant labourers and facilitates the enrolment of their children in schools.

Assam

The Assam Skill Development Mission (ASDM) was established to provide skill training and meaningful employment to unemployed youth. ASDM implemented various initiatives, including the Placement Linked Skill Development Training Programme, partnerships with global knowledge partners, the development of a mobile application called DAKSHA, and the establishment of a skill city and skill development courses in schools and colleges. Additionally, ASDM initiated the "Recruit-Train-Deploy" programme to meet industry requirements for skilled manpower.

7.7. Skill development in agriculture using CSR funds

Some of the skilling programmes implemented using CSR funds are:

Krishi Mitra by Mahindra & Mahindra

Mahindra & Mahindra, in association with Swades Foundation, BAIF Development Research Foundation and Dr Panjabrao Deshmukh Krishi Vidyapeeth, launched the Krishi Mitra project. The project is aimed at educating small and marginal farmers in updated techniques in farming. The farmers were trained in effective farming practices including soil health, crop planning, creating model farms with bio-dynamic farming practices. In addition, the project also provided for soil testing, advisory services, drip irrigation, community farming, seed culture farming, agriculture extension services, infrastructure development and capacity building resulting in improvement in agricultural productivity.

'Project Mooo' under HUL's Prabhat Initiative

Project Mooo is a part of 'Prabhat', a community development initiative of Hindustan Unilever Limited (HUL) launched in partnership with UDAY- a skill development organisation partnered with the NSDC. The project provides holistic

AgTech solution aimed to tackle milk deficit, traceability and private extension service through dairy farm management and ICT training.

The project launched in Rajpura, Punjab aimed at creating sustainable livelihoods for farmers by imparting dairy skills to rural farmers, including women dairy farmers. This included setting up of awareness camps on dairy as a sustainable business, providing digital and financial literacy, visiting farmers at their doorsteps with a team of dairy experts and empowering farmers digitally through Mooo Farm management mobile application. The application harnesses the power of big data to offer demand-led extension training to farmers. The analytics-based learning approach follows the 'just-in-time learning' model offering skills and techniques that transform the dairy farm management practices. The app maps the life cycle of each cattle owned by farmers, sends need-based alerts, offers learning video and tools, connects to local experts, and provides benchmarking status among other features.

Farmer support programme by Adani Ports and SEZ Ltd

Adani is engaged in the upliftment of the rural areas through its CSR initiatives. Its farmer support programme focuses on familiarizing the farmers with new methods and techniques of farming through the adoption of new technologies.

Adani Foundation effectively implemented SRI (System of Rice Intensification) in the villages of Tiroda region in Gujarat. It supported farmers by providing the necessary inputs for taking up SRI. The programme achieved its prime objective of promoting organic methods of paddy cultivation. In the Tiroda region, the SRI method helped 1,000 farmers in reducing their cultivation costs by 33 per cent and increasing their crop productivity by 51 per cent. Many farmers were benefitted following the intervention of the Foundation in association with the Animal Husbandry Department for regular vaccinations, first aid, deworming, and infertility treatment drives in the Tiroda region. The Foundation also empowered the farmers by training them in low-water, labour-intensive and organic methods.

Cairn India Ltd

The Cairn India programme is aimed at increasing income of farming community around sites through productivity enhancement of agriculture and livestock in Rajasthan and Gujarat. The project created a cadre of entrepreneurial farmers/local youth (Cairn AgroFellows) graduated to start their own enterprises in dairying, goat farming, mushroom farming and to work as field extension workers.

Ambuja Cement

The organisation has identified the need for climate smart agriculture and diverted a portion of its CSR funds for agricultural development. Ambuja Cement

Foundation (ACF) implements CSR programme in eleven states, focusing on developing water resources, agriculture, health and education sectors, which in turn facilitate the overall development of farmers. The programmes are implemented by establishing partnership with different stakeholders.

The success of Ambuja Cement in water management can be linked to community participation. In each intervention, Ambuja Cement ensured maximum community participation resulting in large scale adoption of water harvesting structures and technologies by people. Further, water availability led to crop diversification, adoption of plantation and income diversification. These technologies also enabled farmers to adapt agriculture to changing climate and risks associated with it. Skilling the farmers formed one of the key strategies adopted by the ACF.

Indian Oil

Considering the fact green fodder availability in AP is around 40 per cent deficit to the requirement, Indian Oil demonstrated a sustainable model of community fodder bank by establishing a model fodder bank in association with NTR College of Veterinary Science in Gannavaram, Hyderabad.

The project also provided skill based training to small and marginal farmers, in which more than 75 per cent are women. The purpose was to increase the income of marginalized farmers by promoting the use of high protein fodder, which improved the quality and quantity of milk. The project included fodder production, processing and fortification, preservation, packing, storage and distribution amongst small and marginal dairy farmers. Around 2,000 tons of green fodder is being produced every year which is stored in the form of silage bales. The sealed bale preserves the fodder for long term use ensuring availability of quality fodder even during summer/ drought season. The project has helped in balancing the ecology and environment. Fodder grass is sequestering the atmospheric CO_2 resulting in reduction of greenhouse gases.

References

Asian Development Bank, Supporting Human capital development in Meghalaya Project, https://www.adb.org>projects

ASAP Additional Skill Acquisition Programme Kerala, https://asapkerala.gov.in

CEEW Andhra Pradesh Community Managed Natural Farming, https://www.ceew.in>sustainable-agriculture-initiatives

MSDE National Apprenticeship Promotion Scheme (NAPS), https://www.msde.gov.in>apprenticeship-training>naps

NITI Aayog The Utkarsh Bangla Scheme empowers West Bengal through skill development, https://www.nitiforstates.gov.in>best-practice-detail

NITI Aayog Bridging Skills to Employment: Bihar's HUNAR programme empowering youth, https://www.nitiforstates.gov.in>best-practice-detail

NITI Aayog Together we can: Livelihood Collectives in Uttarakhand,

https://www.nitiforstates.gov.in>best-practice-detail

NITI Aayog Kaushalya Vardhan Kendra in Gujarat: Bridging the gap between skill development and employment https://www.nitiforstates.gov.in>best-practice-detail

NSDC Maharashtra Agri Skilling Programme, https://skillsip.nsdcindia.org>knowlwdge-products-maharashtra-agri-skilling

Trent Brown, 2020 Vocational Training, Extension and the Changing landscape of Agricultural Education in India, Working Paper No.407, Institute of Economic Growth, Nov 2020

CHAPTER 8

AGRICULTURAL EXTENSION SERVICES AND SKILL DEVELOPMENT

India has made significant progress in agriculture by increasing production and productivity during the last six decades. After independence, the first planned attempt started with the launching of Community Development (CD) Programme in 1952, followed by the National Extension Service (NES) in 1953 and many such programmes. These programmes were able to educate responsive farmers to take up improved methods of farming across the country.

The major initiatives of the Government of India for developing quality manpower for extension services and the role of FPOs and agricultural start-ups in the development of agricultural sector are discussed in this chapter.

8.1. Extension Service Providers

Agricultural Extension, also known as Agricultural Advisory Services play a pivotal role in boosting agricultural productivity, enhancing food security, improving rural livelihoods and promoting agriculture as a business enterprise. Agricultural Extension, per se, has not functioned as a formal skill training programme, resulting in certification. It has always been less formal, but more activity or practice-oriented. The lack of adequate skill training programmes in agriculture render it difficult for young farmers to interpret their agricultural activity as 'skilled'. To this extent, similar to domestic skills which are unpaid or undervalued, agricultural skills remained invisible and 'taken for granted'.

The Doubling of Farmers' Income (DFI) Committee of NITI Aayog, in defining the mandate of Indian Agriculture, describes Agricultural Extension as: "a system of empowering farmers with information, knowledge, technology, skills, risk and farm management practices, across agricultural sub-sectors and along with all aspects of the agricultural supply chain, so as to enable the farmers to realize higher net income from their enterprise on a sustainable basis".

Extension Reforms in India were pilot-tested in 28 districts of seven States during November,1998 to April, 2005 under Innovations in Technology Dissemination (ITD) component of the World Bank funded National Agricultural Technology Project (NATP). This successful experiment served as a basis to launch the Scheme of "Support to State Extension Programmes for Extension Reforms" by the Ministry of Agriculture, GoI to revitalize Agricultural extension in all the States to make the extension system decentralized and demand- driven.

The draft "Policy Framework for Agricultural Extension" (PFAE) of the Ministry of Agriculture, GoI emphasised that the "policy environment will promote private and community driven extension to operate competitively, in roles that complement, supplement, work in partnerships and even substitute for public extension" (DAC, 2000). The policy also emphasised to leverage the potential of agri-entrepreneurs, para professionals, custom hiring service providers, input dealers and extension workers in non- governmental organizations to supplement the effort of public extension functionaries. It was felt that capacity building of both public extension functionaries and private extension service providers have to be strengthened for effective service delivery. In this context, MANAGE and NIAM emerged as two important national institutes for capacity building of extension personnel in the country.

National Institute of Agricultural Extension Management (MANAGE)

MANAGE responds effectively and pro-actively to the challenges of agricultural extension in a rapidly growing and diverse agricultural sector, which has earned the Institute an international fame and glory. The aim of the institute is to instil managerial and technical skills to Extension officers, managers, scientists and administrators in the field of agriculture and to provide support and services to farmers and fishers for practicing sustainable agriculture.

MANAGE offers its services in the five streams, viz., Management Training, Consultancy, Management Education, Research and Information Services. MANAGE serve as the National Training Partner in Agri Entrepreneurship under Agri Clinics and Agri Business Centres (AC& ABC) scheme of the GoI. MANAGE organises two programmes, viz., Startup Agripreneur- ship Incubation Programme (SAIP) and Agripreneurship Orientation Programme (AOP) for agri entrepreneurs. The institute undertakes several training and extension programmes. Some of the important ones are :

A. Certified Farm Advisor/ Certified Livestock Advisor

This programme is mainly intended to develop/ transform Agricultural Extension Personnel into Specialists. in a particular Crop / Livestock enterprise. The programme consists of three modules. The first module is on basic aspects of latest technologies in Crop /Livestock. After completing the training, the candidate is allotted to Mentor Scientist/s from the concerned Research Institute for a period of one year for technical guidance to face the field level challenges. On completion, the successful candidate is declared as 'Certified Farm Advisor' or 'Certified Livestock Advisor' by MANAGE and the technical partner.

B. Agri Clinics and Agri Business Centres (AC-ABC) Scheme

The scheme provides extension services to farmers through setting up of economically viable self-employment ventures. It aims to tap the expertise available in the large pool of Agri- culture graduates, who can set up Agri Clinic or Agri Business Centre and offer professional extension service to farmers on a payment basis. These Centres established by trained professionals are utilized to support agricultural extension services in the state. (Annexure-4)

The MANAGE also provides agripreneurship training for starting Agri Clinics or Agri Business centres through identified Nodal Training Institutions in various states, which also offer handholding support for one year. NABARD offers credit support to Agri-Clinics through Scheduled Banks. MANAGE is also providing start-up training to graduates in Agriculture, or any subject allied to Agriculture like Horticulture, Sericulture, Veterinary Sciences, Forestry, Dairy, Poultry Farming, and Fisheries. On completing the training, the trainees can apply for special start-up loans for venture.

C. Diploma in Agricultural Extension Services for Input Dealers (DAESI)

In order to improve the technical competency of agri-input dealers in agriculture and related matters and to facilitate them to serve the farmers more effectively by acting as para-extension professionals, MANAGE had launched a self-financed one year 'Diploma in Agricultural Extension Services for Input Dealers (DAESI)' during 2003. The programme is spread over a period of 48 weeks, with 40 classroom sessions and 8 field visits. The programme is implemented by MANAGE through SAMETIs in different states. Agribusiness companies can also sponsor candidates/input dealers for the programme.

National Institute of Agricultural Marketing (NIAM)

With the gradual shift of agricultural system from subsistence to commercial one, there is increasing focus on Agripreneurship and Agri-marketing. SAUs and Research Stations give emphasis on improving production and productivity without giving adequate emphasis on marketing. In order to fill this gap, Ch. Charan Singh National Institute of Agricultural Marketing was set up by the GoI at Jaipur in 1988 to offer specialized training, research, education and consultancy in the field of Agricultural marketing. NIAM plays a vital role in expediting the market reform process and availability of quality managers through its mandate. NIAM organise training programmes in the field of agricultural marketing and allied areas for senior and middle level officers from various line departments of State Governments, Co-operatives, Marketing Boards and Agribusiness entrepreneurs.

The Indian Agricultural Marketing Systems are increasingly posing new challenges in respect of each of the components of the system such as- cleaning, grading and standardization, quality certification, packaging and branding, storage, cold store management, transportation, financing, and wholesaling and retailing. NIAM addresses these challenges mainly through effective capacity development programmes.

CCS NIAM has established ten Centres of training such as Centre for ICT and e-NAM, Centre for Public Private Partnership, Centre for Agri Logistics, Centre for Agri Business Management etc. NIAM also provides consultancy in the area of designing markets, conducting feasibility studies, preparation of detailed project report, and development and implementation of business plan for States in the domain of agricultural marketing.

Other Key Training Providers

State Agricultural Management and Extension Training Institutes (SAMETIs)

SAMETI is the state level institutional mechanism established for catering the training and human resource development needs of Extension functionaries. The main mandate of SAMETI is to develop and promote the application of management tools for effective agricultural extension services in the country. There is one SAMETI in each State. The Agricultural Technology Management Agency (ATMA) at the district level is the convergence platform for all the extension activities in the district, which also provides and monitors the availability of quality inputs and services which the farmers need, besides organising need-based trainings.

Krishi Vigyan Kendras (KVKs)

KVKs function under State Agricultural Universities (SAUs), ICAR Institutes, NGOs, Departments of State Governments, and under other educational institutions. Based on the success of KVKs, the Central Government decided to establish one KVK per district. As of March 2025, there are 731 KVKs established in India. (there are two KVKs in very few districts in India).

All KVKs are mandated to function in reducing the time lag between generation of technology at the research institution and its application to the location specific farmer fields for increasing production, productivity and net farm income on a sustained basis with the main focus of training of farmers and extension personnel to update their knowledge and skills in modern agricultural technologies.

Extension Education Institutes (EEIs)

Extension Division of the Department of Agriculture, GoI has four Extension Education Institutes (EEIs) situated in Nilokheri (Haryana), Hyderabad

(Telangana), Anand (Gujarat) and Jorhat (Assam) which cater to the training needs of field extension functionaries working under agriculture and allied departments of States/UTs of respective regions. The main mandate of EEI is to provide in-service training to middle-level extension functionaries. One of the objectives is to acquaint trainees with the sources of useful information and materials related to their professional needs and help them to develop soft skills.

Schemes under Ministry of Agriculture and Farmers' Welfare

All Centrally Sponsored Schemes (CSS) operating under Ministry of Agriculture and Farmer's Welfare are merged into two umbrella Schemes viz. Pradhan Mantri Rashtriya Krishi Vikas Yojana (PM-RKVY), a cafeteria scheme and Krishonnati Yojana (KY). PM-RKVY aims to promote sustainable agriculture, while KY address food security and agricultural self-sufficiency. All components leverage technology to ensure efficient and effective implementation of the various components.

The **PM-RKVY** comprises of the following schemes:

i. Soil Health Management, ii. Rainfed Area Development, iii. Agro Forestry, iv. Paramparagat Krishi Vikas Yojana, v. Agricultural Mechanization including Crop Residue Management, vi. Per Drop More Crop, vii. Crop Diversification Programme, viii. RKVY DPR component, ix. Accelerator Fund for Agri Startups

Krishi Unnati Yojana comprises of the following important Schemes/Sub missions:

Mission for Integrated Development of Horticulture, National Food Security Mission, National Mission on Sustainable Agriculture, Sub mission on Agricultural Extension, Sub mission on Seeds and Planting materials, Sub mission on Agricultural mechanisation and Scheme Integrated scheme on Agricultural marketing.

Farmers' skill trainings and field extension included in all the components of PM- RKVY and Krishi Unnati Yojana are being converged with similar farmer-related activities conducted by ATMA at the district level.

A variety of new actors entered the space of agricultural education, significantly diversifying the institutional landscape. NGOs have played more prominent roles in agricultural training, particularly in developing skills that they regard as having social and environmental value, such as in sustainable farming (Brown, 2018). Private sector organisations are also more directly involved in the provision of agricultural training, though questions have been raised on their efficacy.

Currently, there are two main systems for public extension services in India. The first comes under the state-level Departments of Agriculture, sometimes referred to as 'frontline extension.' The second public system is overseen by ICAR, which functions as a relatively autonomous organ of the central government. This institution co-ordinates a system wherein extension programmes, largely consisting of short-duration trainings, are administered by agricultural research stations, agricultural universities (each of which has a Directorate of Extension) and Krishi Vigyan Kendras (Agricultural Science Centres).

8. 2. Farmer Producer Organisations (FPOs)

Farmer Producer Organization is a broader term, which includes Farmer Producer Companies (FPCs), farmer cooperatives and farmer societies. The term Farmer Producer Companies (FPCs) refers to producer companies engaged in agriculture and allied activities.

The objective of the scheme is to provide small and marginal farmers 'better collective strength for better access to quality inputs, technology, credit and better marketing access through economies of scale for better realisation of income'.

Some of the important features of FPOs are:

- It is formed by a group of producers for either farm or non-farm activities.
- It is a registered body and a legal entity.
- Producers are shareholders in the organization.
- It deals with business activities related to the primary produce/ product.
- It works for the benefit of the member producers.
- A part of the profit is shared amongst the producers.
- Rest of the surplus is added to its owned funds for business expansion.

Important activities of FPO

It is expected that the primary producers will have skill and expertise in production technology. However, they generally need support for marketing of what they produce, which is mostly absent. The FPOs basically bridge this gap. The FPO will take over the responsibility of any one or more activities in the value chain of the produce right from procurement of raw material to delivery of the final product at the ultimate consumers' doorstep.

Aggregating producers into collectives is one of the best mechanisms to improve access of small producers to investment, technology and market. The facilitating agency has to keep in mind the following factors while organizing FPOs.

- Types of small scale producers in the target area, volume of production, socio-economic status, marketing arrangement.
- Sufficient demand in the existing market to absorb the additional production without significantly affecting the prices.
- Willingness of producers to invest and adopt new technology, if identified, to increase productivity or quality of produce.
- Challenges in the market chain and market environment.
- Vulnerability of the market to shocks, trends and seasonality.

- Previous experience of collective action (of any kind) in the community.
- Key commodities, processed products or semi-finished goods demanded by major retailers or processing companies in the surrounding areas/ districts.
- Support from Government Departments, NGOs, specialist support agencies and private companies for enterprise development.
- Incentives for members (also disincentives) for joining the FPO.

Value chain development by FPOs

Value chains act as an important catalyst in improving farmers' income by strengthening the backward and forward linkages of agriculture. This approach has the potential to increase agricultural productivity, household welfare and build social capital.

Under value chains, thousands of scattered small farms are systematically aggregated and provided centralized services around production, post- harvest, and marketing. This helps reduce transaction costs of the farms for accessing the value chains and makes it easier for small farmers to access inputs, technology, and the market. It also opens opportunities to bring primary processing facilities closer to the farm gates and help producers gather market intelligence and manage the value chain better with digital agriculture tools.

The following measures are necessary for the success of Agricultural Value Chain (AVC).

- Aggregation- include ensuring uniform quality, sorting and grading technology and appropriate storage needs to be resolved to infuse efficiency in processing and marketing.
- Promote good practices and access to timely information to improve bargaining power of farmers vis-à- vis buyers.
- Address governance issues in agri- food system by giving emphasis on product differentiation, food safety, and product standards required in a competitive market environment.
- Encourage private sector engagement through public-private partnerships for enabling farmers to move up in the value chain.

Capacity Development of FPOs

BIRD, NIRD, NIAM and MANAGE are designated as Nodal Agency for knowledge management and training/ capacity building of FPOs on a regular basis.

BIRD: Capacity Building of FPOs is one of the focus areas of Bankers Institute of Rural Development (BIRD) located in Lucknow. It is a Learning Institution promoted by NABARD to cater to the training requirements of the stake holders of agriculture and rural banking. BIRD as an institute of excellence is committed to provide quality training, consultancy and other related services to financial institutions and other stakeholders including international clientele focused on agriculture and rural development.

BIRD has designed six training programmes based on the need, focusing on the entire gamut of activities related to FPOs. They are: Basic Programme on FPOs; Write shop on Developing Business Plans for FPOs; Programme on Financing of FPOs; Programme on FPOs for Trainers (Training Establishments of Banks); Programme on Onboarding of FPOs with Commodity Markets, and Certification Programme on FPO Management. The Programmes are either sponsored by NABARD or customized (State governments, SFAC, Commodity Boards, NCDEX, and IFAD).

Certification Programme on FPO Management is of three months duration with focus on building skills and enhancing the capacities of aspiring CEOs in FPO management.

FPO Academy, MANAGE : MANAGE has setup an FPO Academy with focus on capacity building of different FPO stakeholders and also to conduct research studies, policy advocacy and other related aspects to help government achieve the ambitious target of creating 10,000 farmer-organisations in five years. The Academy is envisaged to function in Public Private Partnership mode. The Academy partners with different public and private institutes for execution of different activities of the Academy, mainly capacity building of different stakeholders. The Academy also seek association of different private institutes as Knowledge Partners covering agribusiness companies, corporates, NGOs, state level federations, etc.

Centre of Excellence (CoE) for FPOs, Karnataka: The Government of Karnataka has established a CoE FPO located in the University of Horticulture Science campus at GKVK, Bengaluru to function as an expert knowledge partner for FPOs and co-ordinate policies and programmes for sustainable FPO management. Three Responsibility Centres have been identified- Organizational capital, Business Capability development and Integrative Technical Capacity. Training and Certification forms one of the important functions of the Centre.

8. 3. Agripreneurs

Agriculture today faces many challenges, and many of these directly or indirectly lead to changing markets, which create both opportunities and risks for farmers. With a growing recognition of the important role of smallholder agriculture for economic growth and rural development in many countries, market-oriented agriculture has emerged an important agenda. Agripreneurship is key in this regard.

Agripreneurship refers to entrepreneurship in the field of agriculture and allied activities. The major agripreneurship opportunities are in agricultural inputs, farming processes, and technology, value chain, agriculture output processing, and in allied sectors like dairy development, poultry, fisheries and so on. Activities like diversification, precision farming, high-tech agriculture, organic farming, etc. are also enhancing sustainable value.

The entrepreneur can create new combinations of production factors such as new methods of production, new products, new markets, identifies new sources of supply and new organizational forms; or as a person who is willing to take risks; or a person who, by exploiting market opportunities, eliminates disequilibrium between aggregate supply and aggregate demand, or as one who owns and operates a business.

Capacity building of Agripreneurs

Khadi and Village Industries Commission (KVIC), Micro, Small, and Medium Enterprises Development Organization (MSME-DO), National Small Industries Corporation Limited (NSIC), National Institute for Entrepreneurship and Small Business Development (NIESBUD), Centre for Entrepreneurship Development, Indian Institute of Entrepreneurship (IIE), State Directorate of Industries (SDIs), District Industry Centres (DICs), Industry Associations are actively involved in the capacity building of entrepreneurs in India.

8.4. Agri Startups

A startup is a young company founded by one or more entrepreneurs to develop a unique product or service and bring it to market. Startups can provide missing links in the agri value chain and deliver efficient products, technologies and services to the farmers on one hand and the consumers on the other. From ICT apps to farm automation and from weather forecasting to use of drone and from inputs retailing and custom hiring to online marketing, and from poultry and dairy ventures to smart agriculture and from protected cultivation to innovative food processing and packaging, there are over 300 Agri-Tech startups in India across the value chain, leveraging the use of technology and innovation in business models to impact the large agricultural sector in India.

Agri startups can provide a meaningful solution across the agricultural value chain and can be in the form of a product, a service or an application. The ecosystem for successful startups typically starts with proper mentorship with a mix of technological and managerial support. The broader agri startup ecosystem includes a number of think tanks, research labs, incubators and accelerators.

Aqgromalin is one of the fastest-growing farm diversification companies aimed to supplement the income of small landholding farmers through ready to implement micro-farms. Backed by technology, the full-stack company offers innovative growth ideas to farmers in the areas of Agriculture, Aquaculture and Animal Husbandry. The company adopts a conventional approach that goes beyond the traditional agriculture infrastructure to make the practices affordable and accessible to farmers.

Aqgromalin integrates new-age technologies like Artificial Intelligence and IoT to help the farmers sustain in a multifunctional economy and strict market reforms. The micro-farms can be started in addition to the existing operations in the crop, with minimum space and labour requirement. The micro-farm can be managed by the farmers' family that adds to household income. Through the micro-farm set-up, a few farmers have recorded an increase by 100 per cent in their income.

Samhitha Crop Care Clinics is pioneering the precision farming advisory in India, aiming to empower farmers to adopt sustainable practices to improve crop production and farmer's profitability. Incepted in 2021, the start-up envisions to help farmers grow more and grow healthy produce through sustainable practices. By harnessing a hybrid framework of data, digital, and crop expertise, Samhitha ensures the intensification of crop production and productivity. The team of Crop Doctors provide precision advisory at the individual tree level by connecting data points captured from digital-tagged trees, aerial scans using drones, wireless sensors, and regular field monitoring.

Start up India is a flagship initiative of the Government of India, which aims to build a strong ecosystem for nurturing innovation and start-ups in the country, to drive sustainable economic growth and generate large-scale employment opportunities. The Ministry of Agriculture in partnership with Startup India launched the Agriculture Grand Challenge in 2017. The challenge was a unique opportunity for Agri-tech start- ups with a commercially viable solution to solve for innovative challenges in the sector.

Centre for Cellular And Molecular Platforms (**C-CAMP**) is an initiative of Dept of Biotechnology, Ministry of Science and Technology, Govt. of India. C-CAMP's Centre of Excellence for Agri Innovation, with support from Ktech (Karnataka Innovation and Technology Society) has unveiled the Agriculture Grand Challenge-2 during 2024. The challenges are categorised into two groups: agriculture and allied. In the agriculture sector, the challenge focused on developing technology to enhance shelf life and reduce post-harvest losses.

Incubators and Accelerators for Agri Startups

A startup incubator is a collaborative programme designed to help new startups succeed. The purpose of a startup incubator is to help entrepreneurs grow their business. Startup incubators are usually non-profit organizations, usually run by both public and private entities. Startup accelerators are a type of a business support programme that provide funds, training and mentorship to startups to help them function successfully through the early stages of business growth.

An incubator helps entrepreneurs flesh out business ideas while accelerators expedite growth of existing companies with a minimum viable product (MVP).

National Academy of Agricultural Research and Management (NAARM) has set up an agri incubation centre in its campus in Hyderabad captioned 'a-IDEA'

(Association for Innovative Development of Entrepreneurship in Agriculture) with the support of Department of Science and Technology, GoI. Inspired by the success, a-IDEA has launched a relatively more ambitious Phase-II of the programme titled 'Agri Udaan'. It is India's first Food & Agribusiness accelerator organised by NAARM in association with IIM-A, CIIE in partnership with Caspian Impact Investment and supported by DST. Centre for Innovation Incubation and Entrepreneurship (CIIE) located in IIM, Ahmedabad is a collective of interventions in the space of innovation-driven entrepreneurship in India.

Innovation Hub (iHub) was launched by ICRISAT to support agricultural tech entrepreneurs, scientists and technology experts who can collaborate to innovate cutting edge ideas across the whole agriculture value chain.

The MANAGE Incubation Centre under AC-ABC scheme focuses development of startups in the major areas like Agri input, ICT in Agriculture, Animal Husbandry, Nutrition and Health, Farmer Service Centre, Post-Harvest Technology, Farm Mechanization, Supply Chain Management, Fishery and Warehouse Management.

8.5. Skill Development by the public sector

The MoA &FW through the Department of Agriculture and Farmers' Welfare (DAC&FW) and the Department of Agricultural Research and Education (DARE) have undertaken important initiatives towards ensuring a greater outreach of extension and education services to individuals aspiring for a career in agriculture and allied sectors. The Indian Council for Agricultural Research (ICAR) leads agricultural research and education. ICAR primarily caters to building professional capacities capable of undertaking advanced technical research and ensuring its effective outreach and adoption among the farmers.

8.1 NAARM (National Academy of Agricultural Research Management)

NAARM plays a crucial role in skill development, particularly in the agricultural sector, by organizing training programmes and capacity building workshops focused on agricultural research and education management, as well as agribusiness entrepreneurship. The key areas of NAARM's skill development initiatives are:

- Training programmes to impart knowledge and skills related to entrepreneurship and innovation in the agribusiness sector, targeting agri-graduates and students.
- Initiated postgraduate education programmes and established an Agribusiness Knowledge Centre to integrate agriculture with agribusiness and raise rural incomes.
- Organizes skill development training on integrated farming systems to help farmers diversify their income sources.

- Provides handholding support to selected State Agricultural Universities (SAUs) to create awareness among agri-graduates for entrepreneurship development.
- Emphasizes the growing importance of Artificial Intelligence and ICT tools for applications in agribusiness.

Source: https://naarm.org.in

The Agricultural Education Division, ICAR is involved in strengthening and streamlining of higher agricultural education system to enhance the quality of human resources in agri-supply chain to meet future challenges in agriculture sector in the country. This calls for regular planning, development, coordination and quality assurance in higher agricultural education in India. The division strives for maintaining and upgrading quality and relevance of higher agricultural education through partnership and efforts of the ICAR-Agricultural Universities (AUs) system comprising of State Agricultural Universities, Deemed-to-be-universities, Central Agricultural University and Central Universities with Agriculture Faculty.

ICAR's National Human Resource Development (HRD) programme focuses on skill development through various initiatives, including training for trainers, strengthening agricultural education, and promoting rural entrepreneurship, with the aim of enhancing the skills of researchers, teachers, and extension specialists. ICAR has formulated an HRM policy to address the training needs of its employees, encompassing model training schemes for scientific, technical, administrative, and skilled supporting staff.

ICAR organizes various skill development training programmes, including those on livestock and poultry production, natural resource management, and banana fibre-based handicrafts. ICAR has developed a manual for trainer's training programs focused on skill and entrepreneurship development in animal husbandry.

Agricultural Extension Division, of ICAR through a network of 11 Agricultural Technology Application Research Institutes (ATARIs) and 731 Krishi Vigyan Kendras (KVKs) delivers agricultural extension and education, which are an integral part of the overall skill development agenda of the Government of India. The major activities of Agricultural Extension Division are technology assessment, demonstration and capacity development.

ICAR-Agricultural Technology Application Research Institutes (ATARIs) upgrade the knowledge and skills of all the staff of KVKs by arranging capacity development programmes. Agricultural Technology Information Centres (ATICs) serve as single window delivery systems and provide technology information, technology services and technology inputs to many farmers.

With the vision to promote innovation and entrepreneurship, ICAR created an Incubation Fund supporting 25 agribusiness incubation (ABI) centres located in various institutes for fostering commercialization of ICAR technologies. These ABIs undertake various activities including business plan development, conferences and seminars, short-term and long-term consultancy; enrolling entrepreneurs in incubation, graduating entrepreneurs from the incubator; and value chain development.

Pandit Deen Dayal Upadhyay Unnat Krishi Shiksha Yojana under Unnat Bharat Abhiyan (UBA) is another promising programme that aims at creating an inclusive academic system wherein higher academic institutions play a strategic role in the rural development mission.

The Ministry of Food Processing Industries (MFPI) is actively involved in ensuring the availability of skilled manpower for the Indian Food Processing Industry. By increasingly focusing on the quality of food, stringent processes need to be adhered to in the processing sector, which requires skilled and trained manpower. MFPI is working with the Food Industry Capacity and Skill Initiative (FICSI), the Sector skill council (SSC) in food processing, to complete the validations of QPs for each job role. It is also involved in the development of course curricula through the National Institute of Food Technology Entrepreneurship and Management (NIFTEM).

Under Pradhan Mantri Kisan SAMPADA Yojana, MFPI supports skilling infrastructure and development of course curricula. The scheme comprises two components: i) development of course curricula for training modules and their translation to English, Hindi and regional languages; and ii) assistance for the creation of infrastructure facilities for skill training centres.

The Indian Institute of Food Processing Technology (IIFPT) and the NIFTEM conduct regular programmes on skill development and entrepreneurship for youth, farmers, self-help groups and industry. MFPI, through NIFTEM and IIFPT, focuses on entrepreneurship development by ways of skill and capacity building, outreach programs and village adoption programs. It has included a module on entrepreneurship in the course curriculum pertaining to all job roles.

In addition to the skill development initiatives undertaken by MSDE, MoAFW and MFPI, prime autonomous public educational institutions like the Indian Institute of Management (IIM) and the Indian Institute of Technology (IIT) have been playing an important role in advancing innovations and entrepreneurship through their respective incubators that support innovations and nurture entrepreneurship. Through the Bharat Innovation Fund, the Centre for Innovation and Entrepreneurship (CIE), affiliated with the IIM-Bengaluru, provides pre-series A and series A funding to game-changing entrepreneurs.

8.2 NSFI

The National Skills Foundation of India (NSFI), a non-profit organization, focuses on skill development and entrepreneurship for enhanced and sustainable outcomes. The NSFI provides training need analysis, training and evaluation, facilitates certification, and provides incubation services. It covers sectors such as agriculture and allied services; water and nutrition; handicrafts; and renewable energy and environment.

NSFI is engaged in building sustainable livelihoods through programme implementation, research and evidence building, policy advocacy and incubation of enterprises for scaled impact. Team NSFI is currently working in 'Farm, Non-Farm, Renewable Energy and Water Management' sectors with a focus on gender equality, incubation & entrepreneurship development, climate action, agriculture-based livelihoods, market development and digital-financial literacy.

Source: https://www.nsfindia.org

8.3 Digital Green

Digital Green, a non-profit organization pioneered the concept of video-based learning in rural India. In 2012, Digital Green partnered with the National Rural Livelihood Mission (NRLM) to enhance the public extension system through its digital approach. The model is based on training frontline workers who screen the videos on best practices related to farming, non-farm practices, financial inclusion and institution building with the larger community.

Digital Green plans to sustain its focus on responsibly integrating AI into the operations. This will involve continuous upskilling, measurement of learning impacts, and the development of policies that ensure safe and effective AI use in all the workstreams that prioritize data privacy and security for the organisation and everyone the organisation work with. By fostering an environment that encourages peer-to-peer knowledge-sharing and skill development, Digital Green aim to bridge the existing skill gaps and propel the team towards higher levels of innovation and efficiency.

Source: https://digitalgreen.org

8.6. Atal Innovation Mission (AIM)

It is Government of India's flagship initiative to create and promote a culture of innovation and entrepreneurship across the length and breadth of the country. The objective of AIM is to develop new programmes and policies for fostering innovation in different sectors of the economy, provide platforms and collaboration opportunities for different stakeholders, and create an umbrella structure to oversee the innovation and entrepreneurship ecosystem of the country.

AIM has launched the Atal Tinkering Lab (ATL) programme, which is a state-of-the-art space established in a school with a goal to foster curiosity and innovation in young minds, between grade 6th to 12th across the country through 21st century tools and technologies such as Internet of Things, 3D printing, rapid prototyping tools, robotics, miniaturized electronics, do-it-yourself kits and many more. The aim is to stimulate a problem-solving innovative mindset within the children of the ATL and nearby communities.

Atal Incubation Centres (AICs) are business incubators established by AIM at universities, institutions and corporates to promote innovation and entrepreneurship among young innovators of the country. These Atal Incubation Centres aim to foster world-class innovation and support dynamic entrepreneurs, who want to build scalable and sustainable enterprises. AICs supports startups from diverse areas such as Healthtech, Fintech, EdTech, Space and Drone Tech, AR/VR, Food Processing, Tourism, among others.

8.7. RSETI

RSETIs are Rural Self Employment Training Institutes, an initiative of Ministry of Rural Development to have dedicated infrastructure in each district of the country to impart training and skill upgradation of rural youth geared towards entrepreneurship development. RSETIs are promoted and managed by commercial banks with active co-operation from the Government of India and State Governments.

One RSETI is established in every district in the country. The concerned lead bank in the district takes responsibility for creating and managing RSETIs. After successful completion of the training, the trainees will be provided with credit linkage assistance by the banks to start their own entrepreneurial ventures.

It is mandated that each RSETI should offer 30 to 40 skill development programmes in a financial year in various avenues. The programmes are of short duration ranging from one to six weeks, Agricultural Programmes include agriculture and allied activities like dairy, poultry, apiculture, horticulture, sericulture, mushroom cultivation, floriculture, fisheries, etc.

RSETIs offer training in 64 vocations classified under four major areas viz. agriculture, processing, product manufacturing and general entrepreneurship development programs. The post training handholding support is provided to the RSETI trained candidates for a period of two years after completion of the training. Also, alumni associations are established by the past trainees in the name of ASARE (Association of Successful Alumni of RUDSETI Trained Entrepreneurs) in many Centres and similar associations are being proposed in other RSETIs also. ASARE provides continuous guidance and support to the trained entrepreneurs for their development and instill confidence and provide a forum for resolving common problems besides enhancing the competencies of members through education and training.

Each RSETI is headed by a Director on deputation from the Sponsor Bank having rural banking exposure with a flair for training. Every RSETI has to have two Faculty Members and support staff. The Director/Faculty Members are exposed to intensive Trainers' Training Program (TTP) organized by National Academy of RUDSETI (NAR) and are enabled to discharge their functions effectively. They also provide with Refreshment training program one year after TTP program to enrich their knowledge towards better functioning of RSETIs.

8.8 Studies on capacity development of Extension service providers

Of late, there are attempts to identify the competencies and skill sets required by professional extension service providers. A sample of the studies in this direction is presented.

Prasad *et al* (2015) reported that Extension Advisory Service (EAS) needs new capacities and skill sets to support farmers efficiently. Skills for facilitation and brokerage in rural development and value chain are very much needed.

Sasidhar and Suvedi (2016) conducted a study on assessment of core competencies of livestock extension professionals in India and reported that extension professionals need a blend of technical skills (disease diagnosis and treatment) and soft skills (communication, problem solving and empathy).

Sorensen *et al* (2021) concluded that understanding of sustainability skills in the context of agriculture as being related and context-dependent is very much needed. The review threw light on how sustainability skills could be assessed, evaluated and operationalized in education and lifelong learning process.

The skill gap report for the Indian fisheries sector (2021) prepared jointly by ASCI and NFDB had identified the major areas for future job creation as: fish feed industry, enclosed culture, recirculatory aquaculture system, square met net fishing and use of digital technologies in fishing. The study has undertaken an exercise to study the trends in employment in fisheries sector and identify the gaps in skills required for the new job roles. The segments covered were aquaculture, mariculture, ornamental fish farming and capture fisheries. It was observed that in India, skill training of the fishers was done mostly through informal, hereditary and learning on the job ways.

Dixit and Ravichandran (2022) reported that currently, the Indian agri food processing sector employs approximately 44 per cent of the workforce, with the majority of the workers lacking any formal or informal skill trainings. Addressing the existing skill gaps between the labour force and industrial demands is essential if the food processing sector is to realise its full potential. Their review presents a realistic picture about the food processing industry, employment generation capacities and skills gaps in the Indian food processing industry.

Agricultural Extension in South Asia (AESA) is a network of individuals and organisations involved in Extension and Advisory Services (EAS) across South Asia. AESA publishes blogs on various topics related to Extension and has published six blogs related to skill development, which are summarised below:

In Blog 8, Prasad (2013) indicated that measures that address the specific skill needs occurring at differing levels of agro-industrialisation have to be initiated. The KVKs established as centres of skill development for the farm youth and practicing farmers need to be properly oriented and steps taken to revamp their functioning.

The report of the National Commission on Farmers (NCF, 2006) recommended for trainings by KVKs in the areas of post-harvest technology, agro-processing and value addition to primary products to provide skilled jobs in villages. The importance of skill development needs stronger articulation and increased funding support.

In Blog 12, based on the experiences of BAIF Development Research Foundation, **Sohani (2013)** illustrated his thoughts on skill development for wage employment, self-employment and enterprise development in rural areas. The skill development programmes should have a rural focus and should offer trainings on farm based and non-farm based trades. It should focus on up-scaling the skills of practitioners and creating a cadre of service providers, or for self- employment.

In Blog 14, Rao (2013) reported that lack of clarity on the different types of skills to be developed and absence of adequate infrastructure for skill development have led to poor skill development among veterinarians and this needs to be changed. He also elaborates the skills required by a veterinarian as a livestock advisor.

In Blog 36, Balaji (2015) had introduced MOOC-for-development as a movement in India and highlighted that MOOCs should be built from the perspective of skills development in agriculture at all levels. MOOC is identified as a channel for youth engagement and the policy environment is India is favourable to build a paradigm of MOOC for development, considering agriculture, food, rural development and well-being of farmers.

In Blog 176, Mahesh Chander (2022) reported that large scale migration adversely affects the rural economy and imposes a resource burden on urban towns and cities. The policies and programmes of the GoI can help rural youth find meaningful occupations in rural areas itself, which would go a long way in productive youth engagement. The various schemes such as PMKVY, DDU-GKY and Skill India campaign have been expanded for a better reach.

In Blog 202, Mahesh Chander (2023) indicated that it is vital to understand the readiness of our youth to embrace the green skills in the agriculture sector. Youth have a huge responsibility to contribute meaningfully by learning green skills and engaging themselves in green enterprises. Promotion of green farming and farm produce from green farming can provide more opportunities for youth trained in green skills. Training of trainers and strategizing the training infrastructure for green skilling are areas that demand attention.

AESA Blogs related to skill development

- BLOG-8: Skilling the farm sector – Need for a focused approach
- BLOG-12: Skill Development for a new Rural Resurgence
- BLOG-14: Skill Development in Veterinary Education
- BLOG-36: Skills, Scale and Speed: Harnessing MOOCs for Large Scale Capacity Development
- Blog 176- Skilling Youth for Sustainable Food Systems Transformation
- Blog 202-Green Skills for Sustainable Development: Is Our Youth Ready?

8.9. Skills Delivery by the private sector

1. Centum WorkSkills India (CWSI)

Centum WorkSkills India Limited (CWSI) is one of the NSDC's largest training partners. It is a subsidiary of Centum Learning Limited, a global provider of customized corporate training programmes and vocational training solutions. CWSI offers an integrated set of services in the skills development sphere, from community mobilization and training delivery to assessment, certification and assistance for placement. It offers an end-to-end skill development solution that follows a structured mechanism, including identification of and outreach to target communities; counselling; training, assessment and certification; and placement support.

CWSI provides vocational training to youth on a large scale. The curriculum designed is based on jobs availability in industry and is validated by industry partners. CWSI selects and certifies training instructors, boosting their career development and lifelong learning opportunities through the Centum Skill Instructors Guild. Mobilization campaigns reach beneficiaries through communities, social specialists, district administrations, self-help groups and other means. After training, youth are placed within industry through a robust industry network. CWSI uses an IT-based system, i-Learner, to manage operations. Innovative cost-cutting measures help operations become more financially sustainable despite low profit margins.

The capacity-building programmes of Centum Foundation for farmers focus on reducing cultivation costs and promoting environmentally sustainable practices. The Foundation offers training and support in agroforestry and establishing green enterprises, enhancing livelihood generation and diversification. The Shram Mitra initiative at the block and village level provides services to the rural workforce, including job availability, upskilling, and social protection.

2. IL & FS Skills

As one of the leading private skilling companies in India, IL& FS Skills provides placement-driven skills training to a diverse range of groups from school dropouts to high school, higher secondary school and ITI graduates. It offers skill development in six categories through its Skill Programme for Inclusive Growth (SPRING) initiative.

The parent company IL & FS established three subsidiary companies, as below:

- IL&FS Education & Technology Services Limited responds to changing needs and requirements in the education sphere as a result of globalization and technological advances.
- IL&FS Cluster Development Initiative Limited identifies barriers to small and medium-sized enterprises (SME) growth such as limited access to infrastructure, recent technologies, financial services and qualified human resources, and offers SME clusters as an innovative solution.
- IL&FS Skills Development Corporation Limited (IL & FS Skills). It builds on the need for effective skills development to utilize the demographic dividend, the demand for a productive workforce and the support of the government to skilling, covering the whole value chain of skills.

The training programmes cover a range of skills required in the manufacturing, agriculture, services, engineering and construction sectors. IL&FS Skills specifically focuses on base-of-the-pyramid groups, from those who have dropped formal education to graduates. The programmes are demand-driven, conducted in cooperation with private sector partners and delivered by trainers in line with recent technologies.

One of the most impactful practices of IL&FS Skills is the development and utilization of interactive multimedia content and the community computer K-Yan. Leveraging technology allows reach on a considerable scale, in terms of the number of people trained, and scope, in terms of the variety of areas on which the trainings are offered. In addition, it ensures quality training delivery even if there are no available skilled trainers or master trainers.

3. IndiaCan Education Private Ltd

IndiaCan, formerly a Joint Venture between Educomp and Pearson, is now fully owned by Pearson. IndiaCan is a pioneering initiative with a view to bridge the gap between employment opportunities, skills and knowledge base of the unemployed youth in the country and helps them get suitable placements.

IndiaCan Education Private Limited responds to industry's need for qualified technical staff and improves the employment prospects and welfare of disadvantaged people, especially unemployed youth.

The key strength of IndiaCan lies in its customized, industry-specific and high-quality content for skill development and vocational training, and a unique business model including effective training of trainers, delivery of life skills and close partnership with the private sector for placement.

IndiaCan draws benefits from the expertise of its parent learning company, Pearson, in designing and instituting training content and practices. The company further adapts training programmes based on the local context. IndiaCan sustains the demand for training programmes by offering courses primarily in retail, hospitality, services, sales and BPOs, which amounts to 80 percent of the market share.

IndiaCan provides placements by partnering with over 400 private companies. Its corporate clients include several major players such as Café Coffee Day, Domino's, Khadim's, Pizza Hut and Reliance.

4. NIIT Yuva Jyoti

NIIT Yuva Jyoti is a skill development venture in association with NSDC that trains disadvantaged youth and school dropouts from cities in India with populations under one million. The organization uses the Source-Train-Place model and seeks 'local jobs for local youth' especially in information technology. 'Source' refers to reaching out to students, understanding their needs and screening them as per industry requirements. 'Train' refers to imparting skills and knowledge assets to students using innovative pedagogy in line with industry needs. The model then moves to match and 'Place' trainees in job roles with ample interview preparation and postplacement (hand-holding) support in the initial period of employment.

The organization leverages 'local jobs for local youth', co-operating with the private sector and small and medium enterprises at the local level to offer training programmes and subsequent employment.

NIIT Yuva Jyoti highlights the skills necessary to react and solve real life problems. It uses Model-Centred Learning Architecture approach along with simulations and case studies, and instructors act as role models for the trainees. Technology also plays an instrumental role in the business model. Innovative tools such as video resumes and online portals facilitate the employment process of the trainees.

5. Basix Academy for Building Lifelong Employability (B-ABLE)

BASIX Academy for Building Lifelong Employability Ltd (B-ABLE) is part of BASIX Social Enterprise group established in 1996 for promoting sustainable livelihoods. B-ABLE, established in 2009, is the first training partner of NSDC for promoting skill building across the nation, and is today one of the key training partners for providing skill and entrepreneurial training in India.

Capacity building of workers in market-aligned job roles and helping them towards an aspirational and dignified livelihood make B-ABLE the most trusted skill development organization. B-ABLE trains workers to be engaged and accountable in their job roles and crafts solutions based on market demand.

B-ABLE developed the company DomesteQ into a social enterprise to provide training and enhance the skills of people working in the informal and unorganized sector, notably empowering illiterate or semiliterate domestic workers who aspire to become drivers, cooks or provide child care or elder care.

Training programmes cover organized services, small business and self-employment-oriented services, industrial and construction skills and green skills including organic farming and eco-tourism. In addition to domain training, B-ABLE recognized a need for soft skill development and developed DISHA, a life skills and behavioural training module designed to help trainees overcome unemployment-related stress, school drop-out, distance from working space to home, and lack of confidence and focus.

6. Dr. Reddy's Foundation (DRF)

DRF is a non-profit partner of Dr. Reddy's Laboratories, a pharmaceutical company. It offers effective short-term skill development programmes that prepare youth aged 18 to 30 for the job market, providing access to market-relevant skills and job placement through its Livelihood Advancement Business School (LABS) model. It aims to "train locally and place locally" to reduce attrition. In rural areas, LABS for Farmers programmes promote transfer of small and appropriate technologies to marginal farmers and, through the SRY programme, undertakes skills development for unemployed youth from farming families.

DRF's interventions provide trainees with market-oriented skills and offer several innovative features, including a livelihood mapping technique that matches demand and supply of skills; work-readiness modules and hand-on learning techniques; and business-to-youth networking for placement opportunities. A strong alumni network and partnership with local companies help to provide local jobs.

7. LabourNet

It is an initiative of Movement for Alternatives in Youth Awareness (MAYA), an NGO. It works with unskilled and semi-skilled unorganized workers by providing skilling and capacity building services, recognition of prior learning, employment linkages, as well as financial inclusion, social protection and welfare services.

LabourNet leverages a model that furnishes end-to-end facilities to vulnerable workers in the informal sector: the 4E model of "Empowerment through Education, Employability and Employment". It links skilled workers with employers, while also providing a legal 'identity' to informal sector workers by offering traceable identity cards, opening the door to social services, insurance, bank accounts and other resources previously unobtainable. The training process includes recognition of prior learning to benchmark the already-acquired skills of workers in the informal sector. The "Learn and Earn" activities offer training and apprenticeship along with placement opportunities. LabourNet has developed an independent Assessment and Certification team, and performs impact evaluations in the form of joint assessments.

8. GMR Varalakshmi Foundation (GMRVF)

GMRVF is a corporate social responsibility initiative of the GMR Group, an infrastructure company. The differentiating aspect of the Foundation's vocational training programme is its tie-up with relevant industries in the provision of sector specific skills development. Industry involvement is an integral component throughout the skilling value chain.

GMR Varalakshmi Foundation undertakes trade specific skill training and entrepreneurship development in a variety of domains. It is equipped with state-of-the-art infrastructure and training facilities, and provides a mix of technical (including hands-on), theoretical and life-skills training. The uniqueness of its skill development programmes lies in its needs assessment of the market and industry before deciding on any course curriculum. GMR Varalakshmi Foundation exemplifies how networking and partnership can positively impact on the economic and social conditions of marginalized youth.

References

ASCI 2021 Skill gap analysis of Indian Fisheries sector, ASCI and NFDB, Nov 2021

Dixit, Preeti and Ravichandran, R 2022 Bridging the skill gaps in the Indian food processing sector: A review, Indian Journal of Research in Food Science and Nutrition, Vol 9, Issue 2, July-Dec 2022

FAO 2017 Information and Communication Technology (ICT) in Agriculture- A Report to the G 20 Agricultural Deputies, Rome

IL & FS Linking Learning to Employability, https://www.ilfsindia.com>media>linking-learning

ISTI Atal Innovation Mission (AIM), https://www.indiascienceandtechnology.gov.in>atal-innovation

MANAGE Diploma in Agricultural Extension Services for Input Dealers (DAESI), https://www.manage.gov.in>daesi>structure

NIRDPR Rural Self Employment Training Institutes, https://nirdpr.org.in>rseti

NITI Aayog 2017 Doubling Farmers' Income (DFI) Vol XI- Empowering the farmers through Extension and Knowledge Dissemination, Ministry of Agriculture, GoI

Prasad, R M, Sulaiman, R .V and Mittal, N 2015 Assessing Capacity Development needs of Extension Advisory Services (EAS): A Review, Working Paper 2015-001, Hyderabad, CRISP

Ravi Nandi and Nedumaran, S 2019 Agricultural Extension System in India: A Meta Analysis, Res Jr of Agri.Sci. 10 (3):473-479

Sasidhar, P V K and Suvedi, M 2016 Assessment of core competencies of Livestock Extension Professionals in India, https://meas.illinois.edu>uploads>2016-07

Sharma, C A., Vinod Kr 2020 The New Age of Agripreneurship, https://www.ramauniversity.ac.in>news>the-new -age-of-culture

Sorensen, Laura Brandit et al 2021 What skills do Agricultural Professionals need in the Transition towards a Sustainable Agriculture? A qualitative literature review, Sustainability 13 (24):

Tripathy, P 2024 How Skill Universities differ from General Universities in India, https://nationalskillsnetwork.in>how-skill-universities

UNDP 2017 How the private sector develops Skills: Lessons from India, https://www.undp.org>istanbul>publication>how-private-sector-develops

World Bank 2015 World Development Report 2015: Mind, Society and Behaviour, Washington DC, World Bank

CHAPTER 9

LIVELIHOOD AND SKILL DEVELOPMENT PROGRAMMES FOR FARM WOMEN

Livelihood development focuses on creating sustainable and diversified income generating opportunities for individuals and communities. The sustainable livelihood is possible only by providing skills for economic growth and inclusive development. Skill development programmes empower individuals by providing knowledge and skills needed to enter into income generating activities.

9.1. Skills development and Livelihood

The programmes of Ministry of Rural Development (MoRD), GoI for self-employment and creation of assets started with Integrated Rural Development Programme (IRDP) in 1980. During the Ninth plan period, Self-Employment Programmes were revamped by merging the Integrated Rural Development Programme (IRDP), the Development of Women and Children in Rural Areas (DWCRA), the Supply of Improved Tool-Kits to Rural Artisans (SITRA), the Training of Rural Youth for Self Employment (TRYSEM), the Ganga Kalyan Yojana (GKY) and the Million Wells Scheme (MWS) into a holistic self-employment scheme called Swarnjayanti Gram Swarozgar Yojana (SGSY). It graduated in 2004 to Swarna jayanti Gram Swarozgar Yojana (SGSY)- Special Projects. The SGSY special projects provided time-bound training and capacity building for bringing a specific number of Below Poverty Line (BPL) families above poverty through skilling and placement in jobs that provided regular wage employment.

To address the challenge of rural poverty, the Ministry of Rural Development conceived a mission mode scheme titled National Rural Livelihood Mission

(NRLM) in 2010. NRLM also known as Aajeevika has been launched by the Ministry to create platforms enabling poor rural people to increase their household income through sustainable livelihood enhancements and improved access to financial services. NRLM was renamed as DAY-NRLM (Deendayal Antyodaya Yojana - National Rural Livelihoods Mission) in March, 2016.

DDU-GKY (Deen Dayal Upadhyaya Grameen Kaushalya Yojana) focuses on skilling and placement of rural youth, while DAY-NRLM (Deendayal Antyodaya Yojana - National Rural Livelihoods Mission) aims to organize rural poor women into Self Help Groups (SHGs) and support their economic activities to improve their livelihoods.

9.2. DDU- GKY

DDU-GKY is a part of the National Rural Livelihood Mission (NRLM), tasked with the dual objectives of adding diversity to the incomes of rural poor families and cater to the career aspirations of rural youth. DDU-GKY is uniquely focused on rural youth between the ages of 15 and 35 years from poor families.

9.1 DDU-GKY

DDU-GKY is a placement-linked Skill Training Initiative which works in Public-Private Partnership (PPP) mode. This scheme makes investments in projects proposed by its training partners, facilitates technical support and provides quality skill training to the poor rural youth. DDUGKY has engagement with Industry heads through workshops and conferences. The industries are invited to be partners with DDU-GKY as Project Implementing Agency (PIA) and/or own a training batch for imparting long-term training for better placement outcomes.

DDU-GKY invests in skill training to make sure that the poor rural youth move out of poverty and achieve prosperity. Funds are invested by DDU-GKY to start advanced, modern and well-equipped training centres for providing private education and skill training. State Governments have a department or State Skill Missions through which they monitor and evaluate these training centres and projects.

DDU-GKY implements training and provides for placement of candidates through building partnerships with Project Implementation Agencies (PIAs), thereby aiding in poverty alleviation. The Training Partners or PIA can register with DDU-GKY through a dedicated Partner Registration process by filing a simple e-Form.

Source: https://ddugky.info

This programme was revisited in the light of the new Skills framework and repositioned under Aajeevika as Aajeevika Skills. It is the skill and placement initiative of the Ministry of Rural Development evolved out of the need to diversify incomes of the rural poor and to cater to the occupational aspirations of the youth.

The focus of the programme is on skilling and placement in the formal sector for rural youth who are poor.

The community institutions provide a collective platform for the rural poor to overcome poverty through access to financial, technical and marketing resources. These institutions require continuous and intensive capacity building and training. The Mission provides multiple funds to these community institutions such as Revolving Fund (RF), Community Investment Fund (CIF) etc. to strengthen their financial base and help them leverage additional funds.

The livelihoods interventions are being supported by Community Livelihoods Resource Persons (Krishi Sakhi, PashuSakhi, Van Sakhi etc.). Custom Hiring Centres (CHCs) are being established at the village level which help small and marginal farmers hire farm tools and services at a nominal cost.

To facilitate intensive implementation, State Rural Livelihoods Missions (SRLMs) have been established as special purpose vehicles. Dedicated implementation support units have been set up at the state, district and block levels known as State Mission Management Unit (SMMU), District Mission Management Unit (DMMU) and Block Mission Management Unit (BMMU). Multidisciplinary professional teams are engaged at each level to spearhead the Mission activities under different components.

9.3. DAY-NRLM

This is the flagship programme of the MoRD for promoting poverty reduction through building strong institutions for the poor, particularly women, and enabling these institutions to access a range of financial services and livelihoods. DAY-NRLM adopts a demand-driven approach, enabling the States to formulate their own State-specific poverty reduction action plans.

Business Correspondent model (BC) has evolved as a crucial vehicle for providing banking services beyond the brick-and-mortar bank branches. Known as BC Sakhi, Business Correspondents are currently providing financial services not only to SHGs and their members but also to the rural households.

The intervention strategy of DAY-NRLM is focused on strengthening the key farm livelihoods of the poor with the following objectives:

- Every poor household is supported on at least two farm sector livelihoods (sustainable agriculture, NTFP, livestock);
- Assured annual incremental income of more than Rs.50,000 per family per year after two years of intervention;
- Yearlong food and nutritional security for the household created by themselves;
- Income streams throughout the year;
- Creation of Social capital for agriculture, NTFP and livestock, from the best practitioners as extension workers;

- The approach has been the creation of knowledge and awareness on livelihoods, initiate sectoral interventions, aggregation and value chain interventions.

DAY-NRLM has been strengthening farm sector livelihoods by empowering the women farmers on adoption of agro-ecological practices, improved livestock management practices, and sustainable Non-Timber Forest Produces (NTFP) collection and value chain interventions.

Under farm livelihoods intervention, the other areas of interventions are Value Chain initiatives and Integrated farming clusters. Value chain development is an important farm livelihood intervention to support small and marginal rural producers to get better market access through aggregation, processing and marketing.

A few newer interventions under farm livelihoods like Integrated farming Cluster, promotion of Sub Sector based cluster interventions, Livestock Cluster were initiated under DAY-NRLM to intensity livelihoods activities.

DAY-NRLM is promoting farm livelihoods through major components/sub-schemes viz.,

1. Mahila Kisan Sashaktikaran Pariyojana (MKSP),
2. Universalization of Farm Livelihoods interventions through SRLM Annual Action Plan;
3. Value chain Development Interventions.
4. Integrated Farming Cluster (IFC).
5. Sub-sector Interventions

9.4. Mahila Kisan Sashaktikaran Pariyojana (MKSP)

Agricultural sector employs 80 per cent of all economically active women, comprising 93 per cent of the agricultural labour force and 48 per cent of self-employed farmers. About 18 per cent of the farm families in India, according to NSSO reports are headed by women. Almost all women in rural India can be considered as farmers in one sense, working as agricultural labour, unpaid workers in the family farming enterprise or combination of the two.

Women in agriculture are generally unable to access extension services and production assets. As most of the women are not recognised as farmers in the absence of ownership rights, they are not treated as beneficiaries of the various government schemes. Due to multiple roles a woman has to perform within the farm and the family, her access to knowledge and information is also constrained. Against this background, to improve the present status of women in farming and to enhance the efforts for their empowerment, Government of India introduced MKSP as a sub component of NRLM.

Value chain development is an important farm livelihood intervention to support small and marginal rural producers to get better market access through aggregation, processing and marketing. The Core strategy of DAY-NRLM in value chain development has been to initiate large-scale interventions and value addition through promotion of Producers' Enterprises (PE) and Producers' Enterprises (PE).

The Producers Groups are informal organizations to support the small and marginal producers at village level through aggregation and marketing in nearby markets whereas the producers' enterprises are formal organizations who support aggregation, processing, value addition and forward marketing. The institutional structure would be based on the opportunities and existing gaps in the value chain.

The Ministry of Rural Development in collaboration with the TATA Trust has set up the FDRVC (Foundation for Development of Rural Value chain) at national level. The Foundation is on a mission to transform the economic potential of small farmers and artisans by incubating producer companies, ensuring they get the best price for their diligent labour and skills.

DAY-NRLM Support Organizations (NSOs) play an important role in bringing domain expertise. DAY-NRLM recognizes select organization of eminence in specific sector/area as NRLM Support Organization to provide technical support to NRLM. The organizations recognized as NSOs under DAY-NRLM are:

- Central Silk Board (CSB) supporting in implementing Tasar-based livelihoods interventions,
- PRADAN supporting in knowledge management, identifying successful applicable models, capacity building,
- Tasar Development Foundation for Tasar based livelihoods promotion,
- NDDB Dairy Services on promotion of milk Producer Companies and
- FDRVC on development and implementing farm value chain development and
- Intellecap on strengthening producer enterprises.

NRETP (National Rural Economic Transformation Project) is the World Bank supported project implemented under DAY-NRLM with an objective to establish efficient and effective institutional platforms of the rural poor to enable them to increase household income through sustainable livelihood enhancements, and improved access to financial and selected public services. The farm livelihoods component of the project has the focus on promotion of producers' collectives, viz. producers' groups and producers' enterprises to enable the small and marginal women farmers to access markets through aggregation and value addition.

9.5. Convergence

- Department of Rural Development has joined hands with Department of Animal Husbandry and Dairying (DAHD) for leveraging SHG platform for rural economic growth through Animal Husbandry and Dairying activities. As per the agreement, it is decided to use the services of Pashu Saki as primary service provider through a new accredited model named as "A-HELP" (Accredited Agent for Health and Extension of Livestock Production).

- Ministry of Rural Development has joined hands with National Highway Authority of India (NHAI) for the promotion of plantations on its lands. NHAI and DAY-NRLM work together for Plantations along the National Highways. SHGs and their federations can take up plantation works along the National Highways.
- Ministry of Rural Development joined hands with the Ministry of Fisheries for providing support in fisheries activities to SHG Members through its Pradhan Mantri Matsya Sampada Yojana (PMMSY). The convergence is aimed at enhancement of production and productivity, Infrastructure, strengthening of safety and security of fishermen or women and providing extension and support system in the fishery sector.
- National Centre for Organic and Natural Farming (NCONF) is associated with the Ministry of Rural Development for promotion of Participatory Guarantee System (PGS) for Organic Certification and Natural Farming Certification.

9.2 Sagarmala-DDU GKY Convergence for Skill Development in Coastal Districts

For skills development of youth in coastal areas, the Ministry of Shipping signed an MoU with the Ministry of Rural Development. The MoU is for convergence of the Ministry's prestigious project Sagarmala and Ministry of Rural Development's flagship programme DDU GKY. This MoU is expected to meet the exact skill needs of industries in the port and maritime sector and provide the coastal communities right skills for better opportunities emanating from the large scale impact of Sagarmala projects. With skill development, there will be close involvement of the coastal communities in Sagarmala and will make them contributors in the economic development and will ensure inclusive growth of the coastal districts of the country. The Coastal Communities are key stakeholders in the Sagarmala Programme and their skill development and livelihood is an important objective of the programme.

Source: https://sagarmala.gov.in>SagarmalaDDUGKY

9.6. Initiatives by SRLMs

Some of the initiatives taken up by the State Rural Livelihood Missions are:

Kerala

In Kerala, Kudumbashree Mission is the agency that promotes and supports the DAY-NRLM activities of the state. It also operates as the National Resource Organisation for the NRLM at the centre.

Agriculture technological fund was launched which extends financial assistance for implementing latest agricultural technologies through the adoption of novel technologies including improved varieties, innovative ideas, innovative crop management system, superior post-harvest processing techniques and value addition units, etc

Value addition fund extends financial support to interested groups for establishing units in various crops like coconut, banana, mango, jack fruit, pine apple, passion fruit etc as part of the value chain activity.

Kudumbashree has ventured into the realm of organic cultivation to promote organic farming and agricultural commodities with organic certification.

The National Rural Economic Transformation Project (NRETP) of National Rural Livelihoods Mission (NRLM) is implemented in many states under the leadership of the Ministry of Rural Development with the support of the World Bank. The aim of the project is to elevate the enterprise model to higher level (scalable enterprises). As a part of this project, NRLM had sought Kudumbashree's support to extend training to the Business Development Support Providers (BDSP), who would be appointed for increasing the business of the enterprises. Kudumbashree had signed an MoU with National Rural Livelihood Mission (NRLM), based on this requirement.

Kudumbashree had extended training to the Community Resource Persons (CRP-EP) of various states as part of Startup Village Entrepreneurship Programme (SVEP). NRLM sought Kudumbashree's expertise to extend training to the BDSPs, as Kudumbashree has previous experience in extending training to the Community Resource Persons (CRP-EP), and the NRETP Project is in the model of the SVEP Project.

'Punarjeevanam' is the Entrepreneurship Development Series of Kudumbashree implemented across the state to revive the livelihood opportunities in the agricultural sector. Punarjeevanam brings about a major change in the agricultural livelihood sector by focusing on skill development required at community level in each district and year-long continuous training and support. Punarjeevanam is a revolutionary step of Kudumbashree in the field of agricultural livelihood.

In the Kerala state budget for 2025-26, a new project titled 'Kudumbashree Livelihood Initiative for Transformation (K-LIFT) is proposed which will ensure livelihood for three lakh women.

Tamil Nadu

Mahalir Thittam is the socio-economic empowerment programme for women implemented by Tamil Nadu Corporation for Development of Women Ltd

(TNCDW), based on Self Help Group (SHG). The hallmark of the SHGs promoted by Mahalir Thittam is the systematic training provided to the SHG members and the office bearers which brings about qualitative changes in the attitude of the women and promotes cohesion and effective functioning of the group.

TNCDW plays a vital role in strengthening the capacity of SHGs and federating them at Panchayat, Block and District level. The focus is on strengthening individual and group enterprises, aggregation of SHG products, value addition, standardisation, and branding for realization of better commercial value by providing online and offline platforms for SHG products to reach local as well as distant markets through skill development and capacity building.

Makkal Katral Maiyams (MaKaMai) were established in all districts registered under Tamil Nadu Societies Registration Act. MaKaMais is an organization that comprise of skilled and experienced community professionals from villages having proven track record of rendering services of implementing poverty reduction programmes. These MaKaMais act as Community Managed Training Centres (CMTCs) for conducting all capacity building training of CBOs.

TNSRLM has initiated the process of engagement of women SHG members as Business Correspondents (BC) for providing basic banking services in the rural areas. Under PRI-CBO Convergence initiative, all SHGs and their Federations have been mandated to prepare Village Prosperity Resilience Plan (VPRP) under four components: i) Entitlement Plan ii) Livelihood Plan iii) Public Goods, Services and Resource Development Plan iv) Social Development Plan

Herbal gardens, millet production clusters, seed production clusters, ornamental fishery and community farm schools were organised under farm livelihood activities.

Community skill schools are set up in all the districts covering 20 members in each batch. The learning is by onsite training methodology by engaging local master practitioners, who have been engaged in running enterprises successfully.

Karnataka

Sanjeevini, Karnataka State Livelihood Promotion Society has made significant achievements in the state over the past decade by creating community based institutions like Self Help Groups, Grama Panchayath and Taluka level federations, various producer groups, Entrepreneur clusters for the rural poor and vulnerable women.

UNDP is providing technical support to the Karnataka State Rural Livelihood Mission (KSRLPS - Sanjeevini) to implement the National Rural Economic Transformation Project (NRETP). Through this project, UNDP is transferring its technical experience gathered over partnership with national and state level programmes. Through this project UNDP is setting up a project management unit, partnerships, knowledge management systems and making innovation to support implementation efforts of KSRLPS- Sanjeevini.

Master Community Resource Person (MCRPs): In a taluk, a pool of 20 CRPs called as Master CRPs are identified and trained to implement the intensive block strategy. They will be divided into 4 teams with 5 MCRPs each who will begin the institution building process in select Village/Ward in the taluk. Each MCRP team goes through one cycle, which is a period of 14 days, covering 2 wards.

Local Community Resource Person (LCRPs): LCRPs are SHG members, who have enough experience as members of SHGs and have developed the faith that SHGs are strong tools for empowerment and poverty alleviation. These successful women have emerged as CRPs to replicate the institutional building process with experiential learning and best practices in our area. All the selected LCRPs undergo five days training by Sanjeevini training team.

Master Book Keeper (MBKs): The Master Book Keeper is responsible for the financial management (maintenance of books of account, audit of SHGs books of account and preparation of financial reports) related duties of the GPLF. Each GPLF will select one Master Book Keeper.

Sanjeevini KSRLPS develops capacity of the MBKs through series of training programmes on maintaining or auditing of financial records and different finance related subjects at GP/block/district level

Telangana

Society for Elimination of Rural Poverty (SERP) is a sensitive and supportive structure to facilitate social mobilisation of poor rural women in the rural areas of all districts in Telangana.

As part of livelihood intervention, SERP has been promoting livestock-based livelihoods under State and Central projects in rural areas with an objective to empower SHG women and to create additional and sustainable income to SHG women with essential backward and forward linkages (Small Ruminant Value Chain). Accordingly, SERP has conducted training programmes on First Aid Veterinary Health services to the interested SHG members with minimum educational qualification 10th class and trained them. Now, SERP has started six days training to SHG women with 60 members per batch on First Aid Veterinary Services.

9.3 Flipkart inks MoU with SERP for market access

In June, 2022, Flipkart, India's homegrown e-commerce marketplace signed a Memorandum of Understanding (MoU) with the Society for Elimination of Rural Poverty (SERP), Department of Rural Development, Government of Telangana, to enable market access for Farmer Producer Organisations (FPOs) and Self-Help Groups (SHGs) in the state of Telangana. Through this partnership, Flipkart aims to empower local farming communities and SHGs of Telangana by providing pan-India market access to more than 400 million customers on its platform. As part of this collaboration, Flipkart will directly source premium quality pulses, millets, staples, and spices from the local farmers, contributing to improved income and business growth.

Source: https://www.businesstoday.in

Maharashtra

Maharashtra State Rural Livelihood Mission (MSRLM) has been constituted and registered by the Government of Maharashtra with an objective to enhance social and economic empowerment of the rural poor in Maharashtra through development of their self-sustained and community managed institutions

The Government of Maharashtra has launched the State of Maharashtra Agribusiness and Rural Transformation (SMART) Project to promote farming in the state. It is a project in association with the World Bank to transform rural Maharashtra using 'SMART' interventions in agriculture and livelihood sectors. The SMART project supports the post-harvest value chain and brings efficiency to benefit the economy at large, encouraging small and medium enterprises within the value chain. It strives to establish a partnership between various stakeholders in the agri-business segments along with the farmer organisations, village level groups, farmer companies, start-ups, SMEs and large corporates with women SHGs. The entrepreneurial partners of the projects include Maharashtra Village Social Transformation Foundation (MVSTF), Maharashtra State Rural Livelihoods Mission (MSRLM) and Maharashtra Agricultural Competitiveness Project (MACP) along with Confederation of Indian Industries (CII).

Odisha

'Mission Shakti' is the self-help mission for empowering women through promotion of Women Self Help Groups (WSHGs) to take up various socio-economic activities launched in Odisha. Mission Shakti has the clear objective of empowering women through gainful activities by providing credit and market linkage. Empowerment of women through WSHGs under Mission Shakti is a flagship programme of the Government of Odisha.

One of the objectives of Mission Shakti has been to provide livelihood support to Women SHGs and their federations through various income generating activities. Women SHGs in Odisha have taken up numerous Income Generating Activities such as Agriculture (Rice, Sugarcane, Onion, Ground Nut etc.), Vegetable cultivation, Organic farming (all kinds of Millets, rice, Turmeric, rajma etc.), and Apiculture.

To unlock women's economic empowerment and rural prosperity through a robust forest economy, Bharti Institute and the Department of Mission Shakti, Government of Odisha have signed a Memorandum of Understanding (MoU). As part of three-year collaboration, the Institute and the Department of Mission

Shakti will work together to build a prosperous and sustainable forest economy in Odisha through women-led community enterprises, starting with the districts of Malkangiri, Koraput, Nabrangpur, Rayagada, Mayurbhanj and Kendujhar.

Along with facilitating technology development and direct market linkage, Bharti Institute will work with Mission Shakti to operationalise security of tenure through Community Forest Resource Rights' (CFR) titles, capacity building, financial inclusion and skill development of Women's Self-Help Groups (WSHGs), so that successful women-led community owned enterprises are established by the end of three years.

9.4 Millet Shakti

Millet Shakti is a collaborative initiative between the Shree Anna Abhiyan and Mission Shakti, aimed at promoting millet entrepreneurship, awareness campaigns, and the consumption of millets through women self-help groups (WSHGs) in Odisha. This joint effort focuses on empowering women entrepreneurs, improving nutritional security, and creating sustainable livelihoods by integrating millet-based food items into both rural and urban areas.

By integrating millet-based food items, the initiative seeks to create sustainable livelihoods for women SHGs. The initiative supports the establishment of Millet Shakti tiffin centres, cafes, and outlets where women SHGs can sell millet-based products. The Odisha government has included millets in state nutrition programs like ICDS, MDM, and PDS.

Source: https://milletsodisha.com>missionshakti

Jharkhand

Small scale broiler farms in Jharkhand represent good investment opportunities, with average returns definitely above 100 percent. In this system, technical knowledge is also required for proper feeding, and management of these birds. This intervention was initiated through some entrepreneur farmers who are willing to invest and also have some risk-taking abilities.

In the case of goat, the broader strategy consists of creating a one-stop support centre or Goat Resource Centre to be nested in Livelihood Service Centre (LSC) of the cluster for small livestock farmer which aims to provide input services, facilitate co-operation between stakeholders to develop integrated value chain by leveraging financial resources from various agencies and also through policy advocacy.

Rajasthan

Rajasthan was one of the first states in the country to set up a Skill Mission, Rajasthan Mission on Livelihoods (RMoL) in 2004. RMoL was reconstituted in February 2014 with inclusion of various independent advisers and professionals.

RMoL is now functioning as an apex policy body for skill development in Rajasthan.

Rajasthan Skill & Livelihoods Development Corporation (RSLDC) is a cent per cent government owned not-for-profit corporation registered under Companies Act, working as the main executing body of various short-term skill training programmes in Rajasthan.

Rajasthan Grameen Aajeevika Vikas Parishad (RGAVP) is an autonomous society established by the Government of Rajasthan under the administrative control of Department of Rural Development. It is mandated to implement all rural livelihoods programmes associated with SHG based institutional architecture in the state. Global Rajasthan Agritech Meet 2016 (GRAM 2016), an agri-technology and business event, was successfully used as a platform to transform and catalyse the process of improving the sustainability and economic viability of agriculture and allied activities shaping globalized, market-driven and profitable ventures.

Bhartiya Skill Development University, Rajasthan Centre of Advanced Technology (R-CAT) and International Institute of Advanced Agriculture Skill Development (IIAASD) are prominent institutions in the domain of skilling. IIAASD offers 16 different kinds of certificate courses for Students, Agri-entrepreneurs, Agri Professional and Farmers.

Himachal Pradesh

In a significant move aimed at boosting the holistic development of agriculture and allied sectors in Himachal Pradesh, an MoU was signed in August 2024 between the Himachal Pradesh Kaushal Vikas Nigam (HPKVN) and the Himachal Pradesh Crop Diversification Promotion Project, Phase-II, which is funded by JICA (Japan International Cooperation Agency).

This collaboration is designed to facilitate the participation of young individuals from Himachal Pradesh in the Technical Intern Training Program (TITP) and the Specified Skilled Worker (SSW) programme. Through these initiatives, technical skills, technology, and knowledge related to advanced agricultural practices from Japan will be transferred to India. The aim is to develop a skilled workforce in Himachal Pradesh by providing on-the-job training in various agricultural units and farms in Japan.

The Government of Himachal Pradesh availed a loan from ADB to implement Himachal Pradesh Skill Development Project (HPSDP) in the state to modernize and reform Himachal Pradesh's TVET programmes and scale up training capacity with the aim to improve livelihood opportunities and employability of Himachali youth aged 16 to 30 years.

An Indigenous Peoples Plan (IPP) has been prepared to ensure the local youth's access to skill training and job opportunities. It describes the suggested mechanisms so that the tribal and gender concerns are incorporated into the project design and receive due attention and support during project implementation. The strategy of incorporating these issues would also demonstrate how the project authorities plan to ensure equitable and fair distribution of project benefits among all tribal populations and others in a culturally responsive manner.

Aagri EDGE (Academy for Agriculture Entrepreneurship Development for Growth and Empowerment) was established in Himachal Pradesh.

West Bengal

Tagore School of Rural Development & Agriculture Management (TSRDAM) started functioning at Kalyani, West Bengal to produce human resources meaningfully in diverse disciplines of Agriculture. TSRDAM is an Institute under Tagore Education and Rural Development Trust. This is the first private Diploma Agriculture Institute in West Bengal. The college has blossomed into a centre of excellence and provides a very conducive environment for the learning- teaching process. It provides all the facilities for learning to enable the students to acquire knowledge, skill and ability to face the emerging trends in the agricultural sector.

WBSRLM (West Bengal State Rural Livelihoods Mission), a registered society under the Society Registration Act has been formed to implement the National Rural Livelihood Mission in the state. NRLM has been launched in the state as Anandadhara. The agenda is to mobilize the rural poor and vulnerable people into self- managed, federated institutions and support them for livelihood collectives. In addition, the poor would be facilitated to achieve increased access to their rights, entitlements and public services, diversified risk and better social indicators of empowerment.

WBADMI Project (2012-2019) with the World Bank support was an innovative project to reach out to the small and marginal farmers particularly tribal and women. The Project is unique that, for the first time an integrated effort was made wherein irrigation facilities combined with agriculture, horticulture and fishery activities were created in the command area.

The use of various technologies particularly GIS has revolutionized departmental planning, approval, design and estimation, monitoring and impact assessment of the project. It has helped in more effective decision making, effective, efficient and quality delivery of the desired outputs, judicious use of available resource and cost estimates, saving in costs and wasteful expenses, establishing two way communications between the State and District teams.

References

Business Today 2022 Flipkart inks MoU with Telengana's SREP to enable market access for FPOs, SHG- https://www.businesstody.in>Latest>Deals

Caritas India, Livelihood and Skill Development, https://caritasindia.org>livelihood-skill-development

CSRBox State of Maharashtra's Agri-business and Rural Transformation (SMART) Project, https://csrbox.org>india_organization_project_Maharashtra

IBEF Mission Shakti Yojana, https://www.ibef.org>government-schemes>mission-shakti

Kumaraswamy, C 2020 Convergence under National Rural Livelihood Missions-Special Initiative, Journal of Research in Humanities and Social Sciences, 8 (12): 33-37

MKSP Farm Livelihood, https://mksp.gov.in

MKSP Transforming Agriculture Touching Lives, https://mkspgov.in>MKSP_Handbook_English

MSDE Skills Acquisition and Knowledge Awareness for Livelihood Promotion (SANKALP) https://www.msde.gov.in>other-schemes-and-initiatives

NSRCEL Swavalambane Karnataka Programme, https://nsrcel.org>swavalambane

Raghavan, V P Social Action, Gender equity and Empowerment: The Case of Kudumbashree Project in Kerala, https://www.igidr.ac.in>conf>money>mfc

CHAPTER 10

TRADITIONAL SKILLS: RECOGNITION OF PRIOR LEARNING (RPL)

Recognition of Prior Learning (RPL) is a formal process that evaluates and validates a person's existing knowledge, skills, and experience gained through formal, non-formal, or informal learning, allowing them to gain recognition for these skills. RPL is a skill certification component to enable Indian youth to take on industry relevant skill certification which will help them to secure a better livelihood.

10.1. Berlin Framework for Action

The demand for the recognition, validation and accreditation (RVA) of all forms of learning with a focus on non-formal and informal learning outcomes was expressed through the Berlin Framework for Action, adopted by 144 Delegations of UNESCO member states at the Sixth International Conference on Adult Education (CONFINTEA VI) in Brazil during December, 2009.

10.1 CONFINTEA VI

Action Recommendation for Transformative Adult Learning and Education (ALE) in CONFINTEA VI (2009)

In keeping with our commitment to creating flexible learning pathways within and between types of work, we underscore the importance of recognition of prior learning as well as the validation and accreditation of non-formal and informal learning, wherever appropriate and relevant, to include all sections of the population, particularly disadvantaged and underrepresented groups such as people with disabilities, into open and flexible learning eco- systems. In this regard, specific attention should also be given to including Indigenous communities in all education and lifelong learning processes. Establishing flexible learning pathways is key to allowing mobility between different programmes, levels of studies and sectors of employment, and for learners to choose their learning trajectories according to their talents and interests, taking advantage of the opening up of bridges across education sub-sectors and the labour market.

Source: https://unesdoc.unesco.org

More than 80 per cent of Indian workforce is employed in such informal or unorganised professional sectors, agriculture being the major. In these industries, job qualifications are not strictly laid out. Recognition of Prior learning (RPL) is a possible step to enhance the employability, mobility and income of those unorganised sector workers who have acquired and honed their skills on the job mostly under the guidance of more experienced peers.

11.2. RPL Explained

RPL is a platform to provide recognition to the informal learning or learning through work to get equal acceptance as the formal levels of education.

RPL is a process of assessment of an individual's prior learning to give due importance to learning as an outcome rather than learning as process. RPL is an assessment that acknowledges an individual's existing skills, knowledge and experiences, irrespective of how they were acquired, be it through formal education, non-formal training or informal learning. In the agricultural sector, RPL serves as a pivotal mechanism to validate and certify the competencies of individuals who have gained expertise outside the traditional educational frameworks.

RPL in India is still very nascent. The earliest instance of a skill development programme for existing workers in the unorganized sector is the Modular Employable Scheme - Skills Development (MES-SDIS) conceived by Ministry of Labour and Employment and operationalised since 2007.

Another such initiative is the Skill Knowledge Provider (SKP) scheme propagated by All India Council of Technical Education (AICTE) and operationalized in 2013. The scheme provides for full mobility between formal and vocational streams of education and the job market with multi point entry and exit.

A separate RPL initiative pursuant to the launch of NVEQF was also taken up by Ministry of Human Resource Development via National Institute of Open Schooling (NIOS). NVEQF is designed to support attainment of qualifications through diverse routes - informal learning being one of them. NIOS designed a holistic national RPL framework to assess the competencies acquired by the learner through prior formal and informal experiences and culminating in NVEQF awards. (The Ministry of Labour and Employment developed the National Vocational Qualifications Framework (NVQF), while the Ministry of Human Resource

Development developed the National Vocational Educational Qualification Framework (NVEQF)).

India's VET had no full-fledged functional RPL scheme, where the prior learning of an individual is assessed and recognised. It was only in September 2012, the Ministry of HRD, Government of India has initiated action on RPL as a part of the National Vocational Education Qualification Framework (NVEQF). The NVEQF supports attainment of qualifications through diverse routes.

NIOS has developed an assessment system and has researched to evolve a process of certification to offer recognition of prior learning to disadvantaged individuals who prematurely opted out of the formal education system because of any socio-economic reason. The process attempts to facilitate integration of formal and informal acquisition of skills and offers a career progression and skill up gradation path to such workforce.

Launched in August 2013, the scheme is being propagated as STAR (Skill Training Assessment and Reward) scheme. The objective of this Scheme was to encourage skill development for youth by providing monetary rewards for successful completion of approved training programmes. The candidates completing skill training and assessment by authorized institutions were awarded a certificate and average INR. 10,000 (Rupees Ten Thousand) per candidate. However, this scheme targeted only formal sector employment. Additionally, requirement of Aadhaar number delayed the process of opening bank accounts and transfer of money. Quality of instruction, curriculum and training methodologies varied in quality in different areas. The scheme mandated assessment and certification bodies be separate from training providers. The assessment and certification were done through the respective SSCs or SSC-approved assessment agencies. In the absence of direct intervention by SSCs, the training providers often manipulated the assessment results.

In 2014, ILO proposed action research on RPL using selected three Skill Councils from sectors that employ significant numbers of informally trained workforce and have also published NOS for a few roles. The selected skill councils are Agriculture Skill Council of India (ASCI), Construction Skill Council of India (CSCI) and Electronics Sector Skill Council of India (ESSCI) under the Project ACE. Tea Plantation worker, Banana plantation worker and Warehouse Operator were three RPL relevant job roles identified for the sector. All these roles are traditionally being performed by unskilled labour or skills acquired from peers on the work- site.

The RPL policies and procedures document defines the responsibilities of the actors involved in RPL implementation, provides guidance that should be followed and implemented in order to build and maintain a robust system of implementation and prior learning assessment, aims at assisting NQA-accredited providers in the process of prior learning recognition and to develop and update policies and guidelines in order to meet RPL criteria.

RPL is built on the principles of the individual right to identify and validate non-formal and informal education, of equal opportunity for all individuals, respect for the right to privacy and individual rights, for the creation of a system that enables access to the identification and recognition of non-formal and informal learning, including adequate quality assurance mechanisms, establishment of system reliability and the establishment of a legitimate system.

11.3. RPL as part of NVEQF

The need for RPL in India arises from the national objective of moving towards a lifelong learning society, in which learners will be enabled to take up learning opportunities at chosen stages throughout their lives. Improvement in quality demands higher standards of service and skill training. In some cases, it also requires conformity to legislations and regulations and could be delivered only by people with recognised skills. RPL enables recognition of achievement from a range of activities using appropriate assessment methodology.

Principles of RPL

- Learner focused: RPL should be a gateway, not a barrier to learning. It should promote positive aspects of an individual's learning experience, as opposed to its deficiency.
- Accessibility: RPL should be an accessible and inclusive process, applicable to all learners at all levels.
- Flexibility: A range of different approaches to RPL in terms of both support and assessment should be encouraged to address the diversity of learner needs, goals and experiences across the different sectors.
- Reliability, Consistency and Transparency: The procedure of assessment should be reliable and consistent as well transparent.
- Clarity of role: The roles and responsibility of learners, assessors, learning providers and receiving institutes should be clearly defined.
- Quality: RPL should be underpinned by a qualitative assurance mechanism.

The implementation and management of RPL is with the framework of the following legislations:

- Article 45 of the Indian Constitution- aims to provide all Indians free and compulsory education for all children up to the age of 14 years.
- Universalisation of secondary education
- National Policy on Skill Development- pre-employment training and life long learning
- NVEQF

11.4. RPL Benchmarking Tools

The following steps need to be taken for development of benchmarking tools.

- Establish whether there are suitable unit of competency or competency standards developed by a recognised awarding and certification body and are recognised on the NVEQF or other relevant qualification framework.
- Each unit of competency is a series of work activities directed towards the achievement of the overall outcome. Elements of competencies provide the RPL benchmark.
- For each unit, there are a number of perforce criteria/indicators that specify the required level of performance. The critical aspect of evidence usually reflects the perforce criteria.

11.5. RPL under PMKVY

RPL is a skill certification component of PMKVY Scheme which focuses on individuals with prior learning experience or skills engaged in the unregulated sectors so that their skills can be assessed and recognized through certification. The objectives are:

- To align the competencies of the pre-existing workforce of the country to the standardized NSQF (National Skill Qualification Framework).
- Enhance the employability and/or entrepreneurial opportunities of an individual.
- To provide opportunities for reducing inequalities based on privileging certain forms of knowledge over others

Under PMKVY 3.0, special focus is given to RPL by recognizing prior competencies of the assessed candidates and provides a certificate and monetary reward on successful completion of assessments. Average monetary reward under RPL would be around Rs.2,200 per candidate.

RPL will help in aligning the competencies of the un-recognized workers of the country with the standardized NSQF framework and provide the much needed recognition, enhancing the career/employability opportunities available to them.

The recognizable certification process of RPL will improve the visibility of the less aspirational/popular job roles amongst the youth of the country, as it will provide a sense of pride and social recognition.

To an extent possible, the RPL would focus on all job roles and special focus may be given to informal job roles depending on the sector by ensuring mobilization of the candidates across these job roles.

Under RPL, the candidate can be certified in two ways:

- Full Qualification: In this scenario, a candidate enrolled under RPL is assessed as per the assessment criteria of the QP-NOS of the selected job role. If the candidate clears competency based assessment under the designated Job Role, candidate is awarded a "Full Qualification" Certificate.

- Partial Qualification NOS Based Certification: Under this scenario, a candidate enrolled under RPL is assessed as per the same QP-NOS. The SSCs have to define the minimum assessment criteria that the trainee has to attain for the QP-NOS in order to get NOS based certificate.

  The Training Partner (TP) or any other authority as prescribed by the Steering Committee will identify and counsel candidates eligible for RPL through mobilization camps, advertisements, etc. The mobilized candidates can be counselled, oriented about the standardized NSQF framework and on the basis of their existing competency will be mapped against the suitable level of the concerned Job role for assessments.

Candidates may be mobilized with the help of SSC's Industry (clusters) connect, associations and MSME connect. These will be in the form of Kaushal Melas, Registration camps in clusters and active marketing through local media such as Radio, Newspapers and hoardings.

Benefits to the candidates:

Help individuals acquire the formal recognition of their qualification which has been acquired by them through informal means or during the course of their work in the organization.

Benefits to the Industry:

If the industry gets its workforce certified under RPL, the industry gets a dual advantage:

- Formal recognition of the skill levels mapped with the QPs & Job roles.
- Open up different paths for workers to upskill & grow through short-term training & bridge courses

11.6.RPL Stages

There are two compulsory stages for RPL: Advise and Assess. The candidates participating in the RPL process must complete both stages. The third stage Audit is applicable for randomly selected RPL-certified candidates as a quality checking measure.

I. Advise: The advisory stage supports, informs and guides the candidate on RPL options available. This process is conducted by an advisor and is intended to help the candidate find the most appropriate NSQF-aligned qualification for assessment.

The role of the advisor is to:

- Help the candidate to understand the RPL process and its benefits
- Observe practical demonstration by the candidate of her/his skills to determine and discuss the most appropriate level and scenario for assessment.

- Assist the RPL candidate to compile evidence in support of a portfolio, if applicable.
- Where appropriate, recommend training courses to the candidate before undertaking the assessment.

II. Assess: This stage involves judging the evidence provided and making an assessment. Such an assessment may result in a full qualification or a part qualification. This assessment process needs to be detailed by the certifying body and approved by the NSQC as part of the qualification file.

Assessors approved by the certifying bodies will conduct the assessment and will evaluate the candidate on his/her claim for recognition following the advisory process. The role of the assessor is to:

- brief, advise and instruct the candidate on the assessment process
- undertake the candidate's assessment
- record assessment results
- make assessment decisions as recommendation to certifying bodies for certification or recommendation for bridge training.

III. Audit: The audit stage is intended to ensure the quality of the overall RPL process, which has to be in line with the Quality Assurance Framework for the technical vocational education and training system in India. In the absence of a specific QA framework, a proposed auditory body will have oversight of all the stakeholders under the RPL process. Once approved, this RPL policy shall become a part of the QA framework.

11.7. Significance of RPL in Agriculture

The agricultural workforce often comprises individuals who have developed their skills through hands-on experiences, familial traditions or community practices. RPL provides these practitioners with formal recognition of their competencies, aligning them with the standard NSQF. This alignment not only enhances their employability, but also opens avenues for further education and professional development.

The benefits of RPL in Agriculture are:

- Formal recognition: Individuals receive government certification for skills acquired outside formal settings, enhancing their professional credibility.
- Enhanced employment opportunities: Certified individuals are better positioned to secure employment, advance in their current roles, given the formal acknowledgement of their skills
- Skill upgradation: There is possibility of upgrading the skills based on the assessment of the skills
- Economic empowerment: Certification enables to obtain higher incomes and enhances the livelihood opportunities.

10.2 GTET

Gram Tarang Employability Training Services Pvt. Ltd (GTET) is a social entrepreneurial initiative in skill training working in largely underdeveloped regions of the country and is committed towards providing young people with high quality vocational education and skill training, relevant and recognized certifications resulting in meaningful employment and successful careers in the organized sector.

GTET is committed to a goal of training one lakh farmers in Odisha on skills for employability and entrepreneurship under 'Project Atal: Recognition of Prior Learning' (RPL).

GTET invited Sustainability Co-creators (SUSCO) to help assess economic impact of its agri-RPL efforts using quantitative methods, beyond the qualitative 'Entry-level Impact Assessment' undertaken by it earlier.

Based on the economic impact assessment results, there is sufficient evidence to proclaim that Project Atal is moving just in the right direction via PMKVY's RPL initiative in improving farmers' agricultural skills, farming practices, production and incomes.

In general, most farmers credit PMKVY's RPL for a progressive shift in their lives. The major findings of the study are:

1. Boost in Production (28%): Farmers trained in job-roles such as maize, mushroom, mango, floriculture, organic grower, solanaceous crops, tuber, cotton and citrus witnessed double-digit growth.
2. Enhanced Food Security (22%): Increase in production has helped improve domestic consumption in nutritious foods e.g. mango, mushroom, maize, pulses, citrus, solanaceous and tubers.
3. Augmented Sales (87%): Owing to knowledge of better farm management and jump in production, farmers had larger produce for sale. Thus, most job-roles posted double-digit growths (and triple digits in some cases). These include pulses, paddy, maize, mango, citrus, floriculture, chilies and vermicompost.
4. Surge in incomes (98%): Due to RPL training, incomes accrued (in double and triple digits) across job roles for those particularly in mango, floriculture, mushroom, vermicompost grower, pulses, tuber, maize, cotton, organic grower, paddy, citrus and solanaceous crops.
5. Default Surge in Production of Primary Crop: Mostly paddy and in some cases maize, tuber and solanaceous crops. Farmers who, prior to training, harvested '4-8' and '>8' quintals witnessed a jump of 17% and 5% respectively.

6. Improved basic quality of life: Around 50% of respondents disclosed that they have been investing their additional incomes under various categories to improve their quality of life. Key among investments were those in 'children's education', 'savings accounts', 'food and groceries', 'household goods', 'medicines', and 'toilet and sanitation'.

Source: https://gramtarang.org.in>uploads>2021/10

The recognition of prior learning enhances upskilling and reskilling pathways by shortening the duration of training and offering more personalised learning pathways to adults. While the majority of recognition systems focus on professional and technical competences for the purpose of entering and progressing in the labour market, the recognition of adults' general competences – i.e. cross-field competences that all individuals need for personal fulfilment and development – is less common.

11.8. LabourNet and RPL

LabourNet is a social enterprise that creates sustainable benefits for school drop outs and workers in the informal sector. They build end to end solutions focussed on bridging the gap between skill demand and supply by initiating industry relevant VET for various age groups, including the unorganised sector. It imparts skills training through its unique model implemented across schools, multi-trade skill training centres and on-site "earn and learn" platforms. LabourNet has focused particularly on the construction sector. It recognises that a fully effective RPL programme depends on the existence of a supportive policy and regulatory environment.

IL&FS Skills, a joint initiative of IL&FS Education & NSDC launched Recognition of Prior Learning in Agriculture sector having Dairy Farmer/ Entrepreneur job role under PMKVY in Bihar. The programme aims to impart skills training for existing dairy workers to align the competencies of the unregulated workforce to the NSQF under PMKVY RPL scheme. RPL focuses on enhancing the career/ employability opportunities of an individual as well as provide alternative routes to greater standards of living through skill enhancement. It also envisages to provide opportunities for reducing inequalities based on certain privileged forms of knowledge over others.

11.9. International attention on skills recognition

The ILO Recommendation No. 195 on Human Resources Development (International Labour Organization 2004) calls on member states to establish a framework for the recognition and certification of skills, including prior learning and previous experience, irrespective of the countries where these were required and how-formally or informally.

In recent years, skills recognition has become increasingly important in policy debates to strengthen skills systems and make them more accessible and inclusive. Given the fact that in most countries, majority of learning is acquired informally (e.g. on the job), to ensure access to lifelong learning opportunities for all, governments increasingly seek ways to "identify, assess and certify a person's knowledge, skills and competencies regardless of how, when or where the learning occurred, against prescribed standards for a part (modular) or full qualification" (International Labour Organization 2015, p. 3).

References

AICTE 2013 National Skill Qualification Framework, https://www.aicte-india.org>NSQF_Notification

ASCI, R P L (Recognition of Prior Learning) https://asci-india.com

DGET 2014 Operational Manual for Skill Development Initiative Scheme (SDIS) based on Modular Employable Skills, Ministry of Labour and Employment, December 2014

European Commission 2017 European Pillar of Social Rights

Harris, J 1999 Ways of seeing the Recognition of Prior Learning (RPL): What contribution can such practices make to social inclusion? Studies in the Education of Adults, Vol 31/2, pp124-139

IndiaFilings 2019 STAR Scheme- Ministry of Skill Development, https://www.indiafilings.com>Learn

IL & FS Skills 2017 R P L (Dairy Farmer or Entrepreneur) Programme by IL & FS Schools in Bihar, https://thedairytimes.com

Michela Meghnagi & Michele Tuccio 2022 The recognition of prior learning: validating general competences, OECD Social, Employment and Migration Working Paper No.270, OECD

MSDE 2015 Skill India Mission: Recognition of Prior Learning (RPL), https://static.pib.gov.in>apr>doc2022242548301

Rothboeck, S., Comyn, P and Banerjee, S P 2018 Role of recognition of prior learning for emerging economies: Learning from a four sector pilot project in India, Journal of Education and Work, 31 (4): 1-15

CHAPTER 11

DIGITAL SKILLS FOR AGRICULTURE AND SKILLS UNDER AI

In the rapidly evolving technological landscape, the demand for talent is determined by mapping digital skills with evolving business needs. With rapid advancements in emerging digital technologies, the nature of jobs and the skills needed to perform them are also changing.

The future of jobs and skills will be shaped by a variety of emerging trends driven by technological advancements, industry 4.0 technologies (cloud computing, artificial intelligence, internet of things, robotics and more), environmental concerns, and evolving workforce dynamics. These trends shall not only transform the way work is executed, but also impact the skills required to thrive in the new economy.

11.1. Disruptive technologies in agriculture

The agri-food system in India is ripe for expanded use of digital innovations, due to several reasons, such as:

- Its large size and complexity, with many stakeholders;
- Its large inefficiencies (unsustainable use of land and water, and significant food loss and waste); and
- Asymmetric access to technologies, knowledge, information, and markets (the vast majority of the 1.2 billion people worldwide not being covered by a broadband-capable network live in rural areas)

The list of technological/digital developments that allow such disruptive innovations in the agricultural sector is enormous and very dynamic, since new things and uses are being discovered every day. Some of the examples are:

- Advances in data collection technologies (in situ and remote sensors)
- Adoption of precision agriculture machinery (digitization and datafication of agriculture)
- Advances in data processing (artificial intelligence, internet of things, cloud computation and increased computing power)
- Advances in encryption, data protection and data sharing technologies, and institutions for data sharing (confidential computing and multi-party computation, synthetic data release and advances in data visualization software)
- Advances in institutions for data sharing (reducing information gaps, information asymmetries, administrative costs (transaction costs) and incentive non-alignment)
- Distributed ledger technologies (DLTs) and smart contracts (improving product traceability and integrity, contract certainty, proof/verification of geographic origin, and compliance with sanitary and phytosanitary requirements)
- Food-sensing technologies
- Digitally enabled marketplaces for agricultural products (e-platforms)
- E-extension services
- The digitization of transactions and payments, and data analytics (credit access and use)

11.2. New Skills for Digital Agriculture

The agri-food sector remains critical for livelihoods and employment. Achieving the UN Sustainable Development Goal of a 'world with zero hunger' by 2030 will require more productive, efficient, sustainable, inclusive, transparent and resilient food systems (FAO, 2017 p. 140). This will require an urgent transformation of the current agri-food system.

There are many challenges in the 'digitalization' of agriculture and food value chain. The transformation must be done carefully in order to avoid an increase of a 'digital divide' between economies and sectors and between those with differing abilities to adopt new technologies (OECD). In emerging economies and rural areas, weak technological infrastructure, high costs of technology, low levels of e-literacy and digital skills, weak regulatory framework and limited access to services result in the risks being left behind in the digitalization process.

Conditions for digital transformation

There are several conditions that shape the digital transformation of agriculture in different contexts:

Basic conditions - the minimum conditions required to use technology include: availability, connectivity, affordability, ICT in education and supportive policies and programmes (e-governance) for digital strategies.

Enabling conditions ('enablers')- are factors that further facilitate the adoption of technologies like use of internet, mobile phones and social media, digital skills and support for agripreneurial and innovation culture (talent development, sprint programmes including hackathons, incubators and accelerator programmes).

'Digital literacy' is critical for using digital technologies. Unlike in many developed countries, where students regularly use advanced technologies and digital skills in their education and day-to-day lives, ICT knowledge and skills lag behind in less developed countries. However, there is now an appreciable change, particularly in the Indian context.

Digitalization creates demand for digital skills and for people who are competent in using digital devices, understanding outputs and developing programmes and applications. This requires not only basic literacy and numeracy but also data handling and communication skills. In populations where such skills are lacking, education must improve quickly; ICT is developing at an incredibly rapid pace and rates of learning must keep up (UNDP, 2015).

To encourage digital agripreneurship, companies need to create pools of digitally-skilled employees. This involves finding potential employees with relevant skills and identifying how they can be attracted and retained, as well as recognizing talent that can be nurtured within the existing employee base and investing in developing digital skills in the existing roles.

Today's agriculture must embrace a digital transformation enabled by connectivity. But Indian agriculture remains less digitized compared with many other industries globally. The past advances were mostly mechanical, in the form of more powerful and efficient machinery, and genetic, in the form of more productive seeds and fertilizers. Now much more sophisticated, digital tools are needed to deliver the next productivity leap. Some already exist to help farmers more efficiently and sustainably use resources, while more advanced ones are in the development stage. These new technologies can upgrade decision making, allowing better risk and variability management to optimize yields and improve economies. When deployed in animal husbandry, they can enhance the well-being of livestock, addressing the growing concerns over animal welfare.

Agricultural Technology (AgriTech) is primarily a combination of (1) hardware for infield and outfield; (2) software and applications; (3) data chains for decision support; (4) processes and assimilation and learning; and (5) monitoring and evaluation.

The agricultural technologies belong to five main categories:

1. Data-connected devices (digital ag) using information and communication technology (ICT), internet and artificial intelligence (AI), primarily driven

by software development. Examples include agricultural digital platforms ('ag platforms') and farming apps, which are driving e-commerce and the servicification of agriculture.

2. Ag biochemistry and biotechnology, harnessing the strengths of biotech and bioengineering. This entails the scientific improvement of plants, animals and microorganisms through innovations such as genetic engineering, tissue culture, vaccines, agrochemicals, etc.
3. Innovative food and farming, which unlock new planting systems and food alternatives. For instance, vertical farming (indoor farming), commonly practised indoors and in urban areas, entails the growing of crops (mostly vegetables) in stacks. Companies such as Aero Farms are leaders in this space. They combine geo-visual sensors with big data analytics and AI to develop technologically efficient crops.
4. Farm robotics and automation drawing on mechanical and electronic engineering coupled with AI. Drones are an example of farm robotics; these unmanned aerial vehicles (UAVs) are used in agriculture for monitoring and data capturing. They are often equipped with sensors, which enable the capturing of images to support crop monitoring.
5. Smart warehousing and logistics use blockchain technology, fleet optimisation software and enterprise resource planning (ERP). For instance, radio frequency identification (RFID) and quick response (QR) codes have been used on agricultural value chains to trace commodities from consumers back to the farm. Blockchains are a more recent innovation, and are decentralised, distributed and public digital ledgers used to record transactions in blocks across all computers.

The adoption of AgriTech depends on the 3Cs:

Cost- purchasing a specific AgriTech product or asset-specific investment necessary to use the product;

Complexity of the embedded information and knowledge of a specific AgriTech; and

Capabilities, the level of skill required by a user to learn how to use the technology.

Capabilities of users of AgriTech relate to the level of skill required to use an AgriTech product, which can be categorised as two broad types: (1) production capabilities- refer to the skills necessary for the efficient operation of a farm/factory with new technologies; and (2) linkage capabilities- refer to the skills needed to decode transmitted information from other organisations, allowing diffusion of technology.

The disruptive potential of AgriTech is on a sliding scale. It can be complementary if it supplements existing products, processes or business models, or a replacement/substitute if it displaces existing practices in a sector or value chain and causes behavioural change that ultimately leads to changes in the underlying norms and culture of society.

11.3. AgriTech for skill acquisition and formalisation of jobs

AgriTech is introducing new skill sets to the agriculture sector in India. As agriculture becomes more digitalised and formalised, technical skills such as the operation of drones, platforms and automated systems will be increasingly sought after. These types of disruption range from frugal, involving low costs but providing new markets and new ways of trading with different countries across the border; to more architectural, complementing processes, but expensive and complex.

AgriTech can increase opportunities for youth and women entrepreneurs. AgriTech has opened up opportunities for educated youth to participate in agriculture by appealing to the 'tech-savvy' nature of young Indian farmers who are currently innovating to add value to agricultural value-chain activities.

Challenges

As disruptions accelerate, the major challenges include:

1. New technologies (e.g. 3D printing, AI, data analytics, robotics, etc.) are rapidly redefining some jobs, eliminating others and creating new ones, resulting in a major need to reskill and upskill workforces.
2. Legislation in most countries has not yet caught up to dealing with the risks inherent in our climate crisis. These legislative changes are inevitable and should be anticipated and planned for.
3. Widespread awareness of social injustice is causing close scrutiny of breaches in ethical boundaries and trust, impacting corporate governance and regulations.
4. The current skill sets of leaders, resistance to change and mindset issues are slowing organizations, making them vulnerable to disruptive forces.
5. Due to the rapid changes in both internal and external environments, there is change in company culture. If left to chance, it will likely devolve as the jobs of many get changed or get eliminated. Broadly held beliefs and assumptions that drive culture will require acceptance of a deliberate design of the desired culture.

The key skill categories like Data Science and Analytics, AI and Automation, Cloud Computing, and Cybersecurity are reshaping how businesses operate and thrive. For instance:

- Generative AI Modelling is emerging as a game-changing tool across industries, unlocking creativity and automation.
- Cloud Computing is not only a cornerstone for business scalability but a burgeoning market set to grow to nearly $2.5 trillion by 2030.

- Cybersecurity has transitioned from being a specialized need to a universal imperative in today's digital-first world.

These skills have been on the rise, and paired with enduring human-centric abilities like customer service, empathy, communication, ethics, and cognitive skills demonstrate how technology and human interaction will continue to coexist and complement one another.

Meanwhile, tech skills focusing on cybersecurity, React web frameworks, and the Internet of Things (IoT) are crucial as digital infrastructure and interconnected systems continue to expand.

11.1 China' Innovation Action Plan

China is pursuing a multipronged approach to build a world-leading AI talent pipeline. One notable initiative is the establishment of the "AI Innovation Action Plan" by China's Ministry of Education, which aims to develop 100 interdisciplinary majors in AI, publish world-class AI textbooks, establish AI schools, research centres and launch national-level open online AI courses. The Ministry of Education also plans to train 5,000 students and 500 teachers in AI within the next five years. This initiative underscores China's commitment to nurturing fresh AI talent and integrating AI education into academic institutions to meet the growing demand for skilled professionals.

Source: https://www.weforum.org>stories

11.4. Digital Agricultural Mission

India's digital revolution has significantly transformed governance and service delivery in recent years by creating digital identities, secured payments and transactions. This progress has paved the way for a thriving digital ecosystem across various sectors, including agriculture. The Digital Agriculture Mission is designed as an umbrella scheme to support various digital agriculture initiatives. These include creating Digital Public Infrastructure (DPI), implementing the Digital General Crop Estimation Survey (DGCES), and supporting IT initiatives by the Central Government, State Governments, and Academic and Research Institutions.

The scheme is built on two foundational pillars: a) Agri Stack and b) Krishi Decision Support System.

AgriStack is designed as a farmer-centric Digital Public Infrastructure (DPI) to streamline services and scheme delivery to farmers. It comprises three key components: a) Farmers' Registry, b) Geo-referenced village maps and c) Crop Sown Registry. A crucial feature of AgriStack is the introduction of a 'Farmer ID', similar to Aadhaar card, serving as a trusted digital identity for farmers. The implementation of AgriStack is progressing through partnerships between the Central and State Governments.

The Krishi Decision Support System (DSS) integrates remote sensing data on crops, soil, weather, and water resources into a comprehensive geospatial system.

The Union Finance Minister announced in the Union Budget 2024-25 that the Government, in partnership with states, will implement Digital Public Infrastructure (DPI) for agriculture over the next three years.

To synchronize the Indian IT industry diaspora with the dynamic global market, NASSCOM in partnership with MeitY launched the **Future Skills Prime (FSP)** platform in November, 2020. This platform facilitates seamless access to both free and paid content, assessments, internships etc enabling learners to attain certification in their chosen skill sets.

Future Skills is designed to encompass over 155 skills spanning across 70 job roles in 10 emerging technologies, including Artificial Intelligence, Blockchain, Big Data Analytics, Cloud Computing, Cyber Security, Internet of Things, Web and Mobile Technology, IT marketing, Robotic Process Automation, Virtual Reality, and 3D Printing. The platform offers over 480 courses across these emerging technologies and professional skills. There are over 250 courses aligned to standards/competencies relevant to market demand to enhance employability. This platform also offers courses from Industry like ER&D and BPM (Curated by Subject matter experts) to train the freshers for employment in their sector with digital skills The goal of the programme, under extension, is to re-skill one million learners by 2027.

11.5 Skills under Generative Artificial Intelligence

The Artificial Intelligence (AI) revolution unleashed by the rapid proliferation of Generative AI (GAI) is set to fundamentally alter the structure of the global economy and labour markets. This holds true for India as much as it does for the rest of the world. Like the earlier industrial and internet revolutions, AI also holds the potential to cause disruptions in some areas of the job market, while augmenting human capabilities in other areas and creating new job roles that did not exist before.

LinkedIn's data indicates that AI will impact every sector of the economy, to varying extents depending on the industry. The very nature of GAI enables it to efficiently undertake time-consuming routine tasks, allowing human workers to improve their productivity, and signaling a shift from a **task-based workforce to a skills-first workforce.**

'AI skills' refer to not just one of the core hard AI skills like machine learning (ML) and natural language processing (NLP), but also complementary soft skills like problem solving, communication, and critical thinking, skills which are inherently human and fundamentally irreplaceable. As GAI becomes adept at executing specific tasks, it is these soft skills that are witnessing a significant rise in demand, with professionals possessing these skills being promoted in large numbers than those with only hard skills. Therefore, to adequately take advantage of the opportunities AI provide, it is necessary for any economy's workforce to be adequately skilled in both hard and soft skills so that they can **balance AI skills with people skills.**

India's workforce is already well-positioned to take advantage of the AI revolution, particularly in hard skills. India has the highest relative AI skills penetration rate globally, as evident from the fact that Indians are three times more likely to use GAI than other nationalities, and accounts for one of the fastest-growing AI talent pools in the world. India's youth, particularly GenZ, is also rapidly upskilling itself to retain a competitive edge in the job market. India's start-up ecosystem has already demonstrated its ability to leverage emerging technologies for novel uses, including AI.

LinkedIn's data also highlights three specific predicaments that need to be overcome to ensure a more equitable and sustained distribution of AI's economic benefits.

First, we are entering an age of continuous skilling, where workers must acquire and enhance their skills rapidly and constantly to keep pace with evolving technologies and avoid becoming obsolete. The majority of these skills necessary in the near future are soft skills that are industry- and sector-agnostic. This will challenge the traditional model of higher education and career progression, where industry-specific hard skill degrees were just enough to ensure economic and career mobility. Linkedin helps professionals find new career opportunities through its Generative AI Professional Certificate. However, for a labour market of India's size and diversity, it is also necessary to develop some baseline core GAI and GAI complementary soft skills across the labour spectrum to fully capture the technology's economic benefits.

Second, the few job roles that GAI is likely to disrupt in the short term are those which require advanced degrees and are generally high-paying. Here, the upper-end of the economic strata will be more affected than the lower-end by technological change, unlike in the past technological revolutions predominantly affecting the lower end of the labour spectrum. This suggests that in the immediate future, India's IT workforce, the backbone of its middle class, is likely to be impacted by increased adoption of GAI, potentially causing some economic hiccups, if not mitigated early.

Third, GAI is likely to asymmetrically impact the labour market along gender lines. Not only are women more likely to be impacted by the change in the nature of jobs that AI will bring, they are also less likely to possess the necessary skill sets to take advantage of new opportunities. While this is not unique to AI and is symptomatic of a larger structural gap in the tech industry, it is a special point of concern for India since female labour force participation in the country has traditionally been lower than in similarly placed economies.

These concerns can be alleviated with specific policy initiatives. For instance, the Skill India Mission could include modules on GAI skills along with critical soft skills to empower the majority of India's workforce to thrive in an AI future. Similarly, programmes that introduce GAI to high school and undergraduate students at an early age would be helpful not only in making them familiarise with the technology and its applications, but also in improving the baseline skill levels of India's future workforce.

A reimagined, flexible education system which emphasises training based on a broad, constantly evolving skills framework and greater industry-academia collaboration would also ensure that the capabilities of the labour force are consistently in line with changing employment requirements. Special attention could be paid on policies designed to enable accessible opportunities for AI skilling of women and marginalised groups to bridge existing economic gaps.

Importantly, GAI itself could be made a mandatory tool of pedagogy and skills training in India and could be leveraged to dramatically reduce the cost of training and knowledge transfer. GAI can also enable quick and accurate language translation, allowing workers to upskill themselves in the language of their choice and overcome the linguistic barrier that currently exists in cases where some degree of proficiency in English is necessary to upskill.

The Deloitte AI Institute teams work with industry leaders to help advance AI innovation that can create advanced growth and optimization strategies for the organizations. The Deloitte's report on 'Artificial Intelligence : Augmenting Human Intelligence' sheds light on the progress made so far in AI in India and across the globe, and the way forward. It also gives a glimpse of the technology's potential impact on industries, and the emerging AI landscape across the private and public sectors.

Fueled by a skilled labour force, entrepreneurial spirit and supportive government policies, the Indian technology sector has achieved global excellence, becoming a driving force behind India's immense growth and development.

To understand the AI market fully, its two components- AI services and AI products are to be reviewed and properly analysed to understand how these are growing and diversifying from traditional services and products.

AI services are typically categorised into the following traditional buckets with aspects of AI interspersed in them:

1. IT services: AI enabled IT consulting, implementation, application development and management, outsourcing, deployment, support and education/training
2. Business services: AI-enabled business consulting and BPO services

The AI software product market can be divided into the following archetypes:

1. AI applications: Collaborative content workflow and management, Enterprise Resource Planning (ERP), supply chain management, production and operations, engineering and Customer Relationship Management (CRM) applications

2. AI platforms: Facilitate the development of AI models and applications, such as intelligent assistants
3. AI systems infrastructure software: Analytics, business intelligence software, data management, integration, orchestration, application development, software quality, lifecycle software and application platforms.
4. AI application development and deployment: Analytics and business intelligence software, data management, integration and orchestration, application development, software quality and life-cycle software and application platforms.

As a major theme within the AI market, GenAI is growing aggressively and could reach 33 percent of the global AI market in 2027. Due to this emergence and other shifts towards AI, the industry is witnessing an escalating demand for sophisticated AI software products and service solutions. This encompasses a range of advisory services and implementations of Large Language Models (LLMs) for generative applications, including summarisation, content creation, video analytics and knowledge assistance through advanced chatbot functionalities and complete value chain transformations using AI.

To use and capitalise on this growth opportunity, the Government of India launched the "IndiaAI Mission" with a budget outlay of INR10,000 crore over the next five years. The IndiaAI mission has planned a comprehensive integrated framework focused on technology (algorithms, LLMs, etc.), infrastructure (GPUs, specialised accelerators, Cloud, HPC, etc.), applications (sectorial, domain, etc.), best practices (ethical/responsible AI), talent (technical and non-technical training to develop/ understand/use/contribute to datasets), enhancing general awareness (sectoral opportunities, threats, etc.) and development of a collaborative AI ecosystem among academia, schools and research.

The Report on National Programme on AI skilling approved by MSDE is given in Annexure-8.

11.2 AI-ready talent pool

The Ministry of Electronics and Information Technology (MeitY) has envisioned multiple K12 and Graduate/Post Graduate level interventions to develop an AI-ready talent pool. The MSDE has also collaborated with tech companies to empower India's workforce with future-ready skills and democratise access to AI skills nationwide.

To ensure competitiveness in the global AI landscape, leading industry players are reimagining a full stack of AI and tech skills across three different levels, emphasized across a three-dimensional lens of AI skills:

- Level 1- Skills of the past: These skills are relevant only for the next three-to-four years but will not significantly propel career growth (such as basic programming, data analysis and fundamentals of AI/ML).
- Level 2- Skills relevant today: These AI skills are pertinent in today's environment (such as advanced ML and deep learning, NLP, RPA and AI model deployment).
- Level 3- Skills of tomorrow: These futuristic AI skills will become prevalent in the next few years (such as AI ethics and compliance, autonomous systems, edge AI, training and fine-tuning AI models and quantum computing for AI).
- In line with these levels, there are five core segments that need to be targeted with different learning paths, ranging from generic to specialised:
- AI informed: Focused on general AI literacy
- AI users: Focused on using AI to improve productivity and be AI first in the respective discipline
- AI integrators: IT workforce integrating AI tools and technologies into existing workflows
- AI builders /developers /architects: Workforce that will develop AI-related software and applications
- AI experts: Top-tier talent focused on new-age research and who have the skill and capability to develop AI models from scratch.

Source: https://indiaai.gov.in

11.3 ADVANTA(I)GE INDIA

Microsoft unveiled a significant skilling initiative in India, aiming to equip 2 million people with AI skills by 2025. The ADVANTA(I)GE INDIA initiative is part of Microsoft's Skills for Jobs programme, which is designed to empower India's workforce with future-ready skills. The initiative is part of Microsoft's broader commitment to accelerate India's AI transformation. The skilling initiative is aligned with the company's responsible AI principles, and training will be delivered in partnership with governments, non-profit and corporate organizations, and communities.

India has a huge opportunity to be a global leader in AI, and creating AI fluency at scale is a critical step in that journey. This initiative aims to propel India into a promising era of AI fluency, empowering citizens across India with the right skills to thrive in the age of AI.

Source: https://www.thehindu.com>Sci.Tech>Technology

The need for AI skilling in India has become crucial for economic sustenance and keeping abreast of global competition. The Government of India has launched many efforts to teach AI skills to its people. The aim is to make the nation a leader in the digital market. Two of the key efforts in the given direction are mainly the National AI Portal, known as INDIAai, and NIELIT. The portal is introduced with the help of the Ministry of Electronics and Information Technology (MeitY), the National e-Governance Division (NeGD), and NASSCOM. The National Institute of Electronics and Information Technology (NIELIT) develops and conducts certification courses in AI. It is India's main hub for all AI work and offers plenty of resources.

The India AI mission is a project to promote understanding of artificial intelligence and encourage a favourable setting for AI creativity and building in India. It includes several tasks and plans to advance the employment and incorporation of AI in various fields. The mission includes activities to train and increase people's AI skills, giving them the needed knowledge and skills.

The INDIAai portal acts as a main hub for accessing details, resources, and news about AI efforts, events, and rules in India. By creating a helpful setting for AI, the portal aims to bring digital change and tech progress in different areas.

11.4 India AI Mission

The IndiaAI Mission aims to build a comprehensive ecosystem that fosters AI innovation by democratizing computing access, enhancing data quality, developing indigenous AI capabilities, attracting top AI talent, enabling industry collaboration, providing startup risk capital, ensuring socially impactful AI projects, and promoting ethical AI. This mission drives responsible and inclusive growth of India's AI ecosystem through following seven pillars. They are:

IndiaAI Compute Capacity, IndiaAI Innovation Centre (IAIC), IndiaAI Datasets Platform, IndiaAI Application Development Initiative, IndiaAI FutureSkills, IndiaAI Startup Financing, and Safe & Trusted AI.

Source: *https://indiaai.gov.in*

India stands out from other developing nations by forming strategic partnerships and pushing for bold projects. A key example is the FutureSkills Prime programme, an effort led by the Ministry of Electronics and Information Technology and NASSCOM to teach new skills and improve existing skills in AI professionals. This programme acts as a business-to-consumer platform focusing on how to use AI in both IT and non-IT fields.

The Natural Languages Translation Mission made by MeitY with the Ministry of Electronics and the Department of Science and Technology shows India's commitment to building a society rich in knowledge. This service uses AI to help with translations between English and various Indian languages, making it easier for more people to access digital content. The opening of the Virtual and Augmented Centre of Entrepreneurship in Bhubaneswar, together with the Government of Odisha and IIIT Bhubaneswar, is meant to create a strong research environment for Virtual Reality and Augmented Reality technologies.

11.6. Use of AI in Agriculture

For agricultural companies, generative AI can gather and crunch even more granular and timely information and analysis that help them improve their products, customer service, and processes and operations to better meet business goals, including sustainability targets.

The sustainable use of resources in agriculture is critical. Generative AI uses large scale data analysis to maximize the use of resources. With the use of historical production data and land topography, AI algorithms build intelligent models that help farmers use resources as efficiently as possible.

In the field of Agriculture, AI holds the promise of driving a food revolution and meeting the increased needs of the people by 2050. It also has the potential to address challenges such as inadequate demand prediction, lack of assured irrigation, and overuse/misuse of pesticides and fertilisers. Some use cases include improvement in crop yield through real time advisory, advanced detection of pest attacks, and prediction of crop prices to inform sowing practices.

GAI has emerged as a transformative technology in various industries, and its impact on agriculture is particularly noteworthy. By leveraging advanced algorithms, machine learning, and big data, generative AI is revolutionizing the way farmers manage crops.

Crop Yield Prediction

Generative AI algorithms can analyze vast amounts of historical data, including climate patterns, soil conditions, and crop characteristics, to accurately predict crop yields. By considering various factors, such as temperature, rainfall, nitrogen levels, and disease prevalence, generative AI models can provide farmers with valuable insights into expected crop yields. This enables farmers to optimize resource allocation, plan harvesting schedules, and make informed decisions regarding crop management strategies.

Disease and Pest Management

Identifying and combating crop diseases and pests is crucial for ensuring healthy and abundant harvests. Generative AI models can analyse diverse datasets, including images, sensor data, and historical records, to detect early signs of diseases or infestations. By employing computer vision techniques, these models can identify subtle changes in plant health, enabling timely intervention. This helps farmers minimize crop losses, reduce pesticide usage, and enhance overall sustainability.

Precision Agriculture

Generative AI is playing a vital role in the implementation of precision agriculture techniques. By integrating data from satellite imagery, drones, and ground sensors, generative AI models can generate detailed maps of crop health, soil moisture levels, and nutrient distribution. This information enables farmers to implement site-specific interventions, such as precisely targeted irrigation, fertilization, and pesticide application. As a result, resources are utilized efficiently, minimizing waste and enhancing productivity

Crop Breeding and Optimization

Generative AI algorithms are being employed to expedite the crop breeding process. By analysing genetic and phenotypic data, these models can identify desirable traits and predict the outcome of crossbreeding experiments. This accelerates the development of high-yielding, disease-resistant, and climate-resilient crop varieties. Generative AI can optimize crop growth parameters, such as lighting, temperature, and humidity, in controlled environment agriculture systems, leading to improved crop quality and yield.

Sustainable Resource Management

Generative AI models facilitate sustainable resource management practices in agriculture. By analysing historical and real-time data, these models can optimize irrigation schedules, minimize water consumption, and reduce the environmental impact of farming operations. Generative AI can also optimize the allocation of fertilizers and pesticides, reducing chemical usage and preventing soil and water contamination.

Optimized Resource Allocation

The sustainable use of resources in agriculture is critical. Generative AI uses large-scale data analysis to maximize the use of resources. With the use of historical production data and land topography, AI algorithms build intelligent models that help farmers use resources as efficiently as possible. This leads to less water use, lower energy use, and increased land productivity. Agriculture becomes a classic example of sustainability and fiscal responsibility by using Generative AI's intelligent resource management.

Enhanced Decision-Making and Predictive Analytics

Farmers now have strong tools for predictive analytics and decision-making, thanks to generative AI. GAI is able to forecast future crop yields, market demand, and even possible disease outbreaks by utilizing machine learning algorithms and historical data. This enables farmers to choose crops, plan planting dates, and develop market strategies based on data.

Reduced Environmental Impact

Reducing the environmental impact of the agriculture business is one of the biggest advantages of generative AI. Generative AI contributes to sustainable farming methods by minimizing the usage of agrochemicals and optimizing resource allocation.

For determining the ideal quantity of water, fertilizer, and pesticides needed for a given crop, generative AI systems can evaluate data pertaining to soil composition, weather patterns, and crop growth. By using a precise method, soil degradation is avoided, chemical discharge is decreased, and resource use is minimized. Furthermore, generative AI can help find eco-friendly substitutes like organic fertilizers and natural pest management techniques.

AI and ML in Dairy Management

AI and ML are redefining cattle health management by predicting and preventing diseases, ultimately ensuring healthier and more productive herds.

AI algorithms analyse data from various sources, including temperature sensors, activity trackers, and even the analysis of cow vocalizations, to detect early signs of illness or discomfort. When an anomaly is detected, AI systems can issue real-time alerts to farmers or veterinarians. These alerts can be accompanied by recommended actions, helping farmers take timely steps to address the issue.

Machine learning plays a pivotal role in genetic enhancement programs, optimizing breeding decisions for improved milk yield and quality. AI-driven systems are optimizing cattle nutrition, ensuring that cows receive the right nutrients for maximum milk production. The infusion of AI, ML, and Block chain has unlocked unprecedented potential in Indian dairy. These technologies have revolutionized cattle health management, genetic enhancement, and supply chain traceability.

11.7 AI Startups in Agriculture

Some of the AI startups in agriculture are:

- **Prospera:** This startup has built a cloud based solution that collects all data available to farmers, and then connects to a device in the field that analyses data and suggests desired results. It works with a variety of sensors and technologies such as computer vision.
- **Blue River Technology:** This combines AI, computer vision and robotics to save costs and reduce the quantity of pesticides. The computer vision defines each plant separately, ML determines how the character of each should be observed. Robotics intelligently control the farm machinery and takes suitable actions.
- **Gramophone:** It uses image recognition skills to help farmers get the right information, and practices to be followed at the right time to get the best harvest possible.
- **Jivabhumi:** It creates smart agricultural marketplace for optimizing the supply and demand for agricultural products, which is often inadequate.

Technologies such as blockchain collect information about products at varied stages of supply chain.

- **Taranis:** Established in 2015, Taranis is an Israeli precision agriculture intelligence platform, which harness advanced computer vision, data science, and deep learning algorithms for effective field monitoring. Taranis' crop intelligence solutions leverage artificial intelligence (AI) and remote imaging technologies to autonomously detect crop threats.
- **Fasal:** Founded in 2018, Fasal, an Indian agri-tech platform by Wolkus Technology Solutions, leverages AI to monitor diverse farming conditions. It employs data science to offer on-farm predictions, delivering insights across iOS, Android, Tablet, and web devices. Using sensor data, Fasal's precision agriculture solution suggests real-time agronomic strategies, enabling water, energy, fertilizers, and pesticides cost savings.
- **Trapview:** Crafted by EFOS, a Slovenian company, Trapview is a platform dedicated to pest insect monitoring and forecasting. Automated electronic traps facilitate efficient, near-maintenance-free tracking of pest insects. Employing sophisticated artificial intelligence-based processing of trap data, Trapview provides users with precise, real-time insights into pest population dynamics.
- **Cropin:** It is a Bengaluru-based startup which provides an AI-based intelligence mechanism to various stakeholders in the agritech ecosystem for sustainable and resilient agriculture. The SmartFarm Plus solution from this startup integrates data from a variety of sources, thus digitising farm data. This includes third-party ERP solutions, manual input via the SmartFarm app, earth observation and satellite-based meteorological data, drones, and other IoT devices, using AI and data analytics to support farmers' decision-making process.
- **DeHaat:** This Gurgaon-based startup provides farmers with access to more than 3,200 agricultural inputs, combined with AI-based customised crop advisory on pest and disease management, delivered via mobile app and call centres. With real-time data, farmers can get weather forecasts beforehand to make arrangements. Moreover, the insights help in risk management and also mitigation.

11.8. NITI Aayog and AI

Automation in agriculture is an emerging subject across the world. In recent time, Artificial Intelligence has been witnessing a lot of direct application in farming. AI-powered solutions not only enable farmers to do more with less, they also improve the quality and ensure faster go-to-market for crops.

Advances in computer vision, mechatronics, artificial intelligence, and machine learning are enabling the development and deployment of remote sensing technologies to identify and manage plants, weeds, pests and diseases. This also provides a unique opportunity to develop intelligent seeding methods for precise fertilization. Artificial intelligence solutions can enable farmers not only to reduce wastage, but also improve quality and ensure faster market access for the produce.

Cognitive computing has become the most disruptive technology in agricultural services as it can learn, understand, and interact with different environments to maximize productivity. Microsoft is currently working with 175 farmers in Andhra Pradesh to provide agricultural, land and fertilizer advisory services. This initiative has already resulted in 30 per cent higher average yield per hectare.

Proximity sensing, remote sensing, Internet of Things (IoT) and image-based Precision Farming are being used for intelligent data integration related to historical meteorology, soil reports, recent research, rainfall, insect infections and along with drone imagery is being used for in-depth field analysis, crop monitoring and field surveys.

AI is a constellation of technologies that enable machines to act with higher levels of intelligence and emulate the human capabilities of sense, comprehend and act.

AI gets categorised in different ways and it is useful to understand the various categories, their rationale and the implications.

(a) **Weak AI vs. Strong AI:** Weak AI describes "simulated" thinking, which means a system which appears to behave intelligently, but does not have any kind of consciousness about what it is doing. For example, a chatbot might appear to hold a natural conversation, but it has no sense of who it is or why it is talking to you. Strong AI describes "actual" thinking. Here, it is behaving intelligently, thinking as human does, with a conscious, subjective mind. For example, when two humans converse, they most likely know exactly who they are, what they are doing, and why.

(b) **Narrow AI vs. General AI:** Narrow AI describes an AI that is limited to a single task or a set number of tasks. For example, the capabilities of IBM's Deep Blue, the chess playing computer that beat world champion Gary Kasparov in 1997, were limited to playing chess. It would not have been able to win a game of tic-tac-toe - or even know how to play. General AI describes an AI which can be used to complete a wide range of tasks in a wide range of environments. As such, it is much closer to human intelligence.

(c) **Superintelligence:** The term "superintelligence" is often used to refer to general and strong AI at the point at which it surpasses human intelligence, if it ever does.

11.9. Challenges in the use of AI

The challenges are concentrated across common themes of:

(a) Lack of enabling data ecosystems

(b) Low intensity of AI research

i. Core research in fundamental technologies

ii. Transforming core research into market applications

(c) Inadequate availability of AI expertise, manpower and skilling opportunities

(d) High resource cost and low awareness for adopting AI in business processes

(e) Unclear privacy, security and ethical regulations

(f) Unattractive Intellectual Property regime to incentivise research and adoption of AI

The funding issues are a large challenge for the progress of AI skilling programmes in India. Some possible solutions to improve financial conditions include finding various financing mechanisms like outcome-based bonds, Social Impact Bonds (SIB), Development Impact Bonds (DIB), and fund-of-funds. Public-private partnerships are also suitable means to tackle these issues.

11.10. Skilling for AI

Re-skilling of the current workforce will require integration with relevant existing skilling initiatives, building of new platforms that can enable improved learning, and novel methods of allowing large scale employment generation through promotion of AI.

Creation of open platforms for learning: Initiatives such as the NASSCOM Future Skills Platform will play an instrumental role in large scale dissemination of requisite skills to some major sections of the employed workforce. Online and self-learning platforms, such as Coursera and edX, are able to connect learners to the best universities and institutions from around the world, and can play a crucial role in this scenario. There is need to bring out guidelines for promoting these while ensuring uniformity, standards and usability. As in the promotion of MOOCs, large scale deployment and adoption of these platforms requires stringent measuring of quality, and recognition of their certification.

AI will be the tipping point in technological evolution of mankind, with human dependence on machines and algorithms for decision making attaining a high level. Thus, any strategy document on promoting AI, necessarily needs to be conscious of the probable factors of the AI ecosystem that may undermine ethical conduct, impinge on one's privacy and undermine the security protocol. Appropriate steps to mitigate these risks need to be an integral part of any such strategy.

There are three types of AI: ANI, AGI, and ASI. The first type, ANI (Artificial Narrow Intelligence) is now existing everywhere, like Google maps, which is highly useful at finding efficient routes to places while riding a vehicle. The second type is AGI (Artificial General Intelligence), which is like a computer, as smart as a human in all aspects. Hence anything humans can do with our brain, it can also do, including learning. AI is getting powerful day by day with applications leading to machines and systems resulting in more advanced AI i.e. ASI (Artificial Super Intelligence). This leads to a stage when a computer or a system becomes better than a human being, wiser, more creative, more socially adept, and this ranges from being a little bit better to being smarter than the sum of all humanity combined.

Natural Language Processing (NLP), Robotics, Machine Learning (ML), Automated Reasoning, Knowledge Representation, Expert Systems, Computer Vision, Speech Recognition, Automated Data Analytics, Virtual Reality, Augmented Reality, Internet of Things (IoT), Cloud Computing, Statistical Computing, Deep Learning etc. are some major sub-areas of AI having huge potential in solving complex problems of agriculture.

NITI Aayog has recently released a discussion paper wherein it envisions AI solutions for key sectors including agriculture. In agriculture, there is a great potential of AI machines to provide information to farmers on the quality of soil, when to sow, where to spray herbicide, and where to expect pest infestations. Thus, if AI systems can advise farmers on best practices, India could see a massive farming revolution. However, such a futuristic scenario has a formidable challenge of scaling up to cover the entire value chain with factors like capacity expansion and cost reduction in mind.

AI can be used to create intelligent systems embedded in machines that can work with higher accuracy and speed than humans and at the same time, be responsive like humans. AI together with Internet of Things and Sensor Technology could be the great enabler of precision agriculture. AI can also play a critical role along with remote sensing technology in wide scale implementation of Climate Smart Agriculture.

Automation, Sensors, Drones, IoT, Solar Power aided with AI provide new opportunities for business and entrepreneurs to deliver innovative solutions as service at affordable prices to the farmers.

At the national level, AI based automatic grading and sorting are already being done for vegetables and fruits with a view of creating an international agri-commodity standard aiding reliable trading across country boundaries. Deep learning and advanced image processing techniques are used for viewing images and pictures thus digitizing food quality.

11.11. Skill India Digital Hub (SIDH)

NSDC facilitates access to a variety of training programmes for trainers and assessors across different sectors and states. The platform offers details about upcoming training schedules and allows users to search for trained professionals.

Government has taken a major step towards digital ecosystem for skilling through the launch of Skill India Digital Hub (SIDH) Platform in September 2023. SIDH is a digital platform to synergize India's skill development, education, employment, and entrepreneurship landscape. Its primary objective is to provide a comprehensive and accessible platform for skill enhancement, offering industry-relevant skill courses, job opportunities, and entrepreneurship support.

SIDH is a digital expansion of skilling opportunities and integrated platform that have digital job exchange between stakeholders. SIDH is at the intersection of two of the most important initiatives of the Government- Skill India & Digital India. Built on Mobile-First approach, SIDH caters diverse needs of a citizen for skill development such as selection of courses, schemes, apprenticeships, and job opportunities based on preferences and aspirations, Learning Management System for digital skilling, Portable Verified Credentials, etc. It aims to serve India's Digital Public Infrastructure (DPI) for skill development, education, employment and entrepreneurship landscape of India.

The SIDH platform offers courses and apprenticeships opportunities to the learners. The platform is envisioned to integrate with MSME and facilitate access to entrepreneurship schemes to the learners and job seekers.

Takshashila, the National Portal for Trainers and Assessors, is an initiative by NSDC under the Skill India Mission. The portal serves as a dedicated online platform for the management of Trainers and Assessors of the Indian Short-Term Skill ecosystem, functioning as the central repository of information concerning development of quality Trainers and Assessors.

Directorate General of Training (DGT) under the aegis of MSDE implemented the Skills Strengthening for Industrial Value Enhancement (STRIVE) project with the objective of improving the relevance and efficiency of skills training provided through ITIs and apprenticeships. Under the project digital content in blended mode was developed for 14 trades in two phases.

Skill India Digital has established itself as a comprehensive platform for skill development and career advancement in India. To further its impact, the programme aims to identify and explore opportunity areas where startups and innovators can contribute, enhancing the platform's effectiveness and reach.

T-Hub, in collaboration with NSDC and HDFC Bank, is launching Skill India Digital Accelerator, to identify and support startups with grants for addressing the challenges of the modern dynamic workforce through innovative technologies. The programme aims to nurture 8-10 promising startups within specific thematic areas by offering unparalleled support and grants. These selected startups will play a pivotal role in advancing the capabilities of the workforce, pushing the boundaries of what is possible within skill development and employment readiness.

11.12. Knowledge Economy

The World Bank views knowledge economy by associating it to a four pillar framework that analyses the rationale of a human capital based economy.

- An educated and skilled labour force: The establishment of a strong knowledge-based economy require workers the ability to continuously learn and apply their skills to build and practice knowledge efficiently.
- A dense and modern information infrastructure: for easy access to the information and communication technology (ICT) resources to overcome the barrier of high transaction cost, and to facilitate the effectiveness in interacting, disseminating and processing the information and knowledge resources.
- An effective innovation system: a great level of innovation within firms, industries, and countries to keep up with the latest global technology and human intelligence to utilize it for the domestic economy.
- Institutional regime that supports incentives for entrepreneurship and the use of knowledge: An economy system should offer incentives for better efficiency in mobilizing and allocating resources, together with encouraging entrepreneurship.

Indian Institute of Information Technology and Management, Kerala (IIITM-K) was established as an autonomous premier educational institution by the Government of Kerala in 2000. To steer Kerala into a leadership role in the emerging knowledge economy, IIITM-K was upgraded to "Kerala University of Digital Sciences, Innovation and Technology" in 2020.

The Centre for Digital Innovations and Product Development (CDIPD) is a Centre of Excellence established by Digital University to innovate, design, develop, and implement digital solutions on multiple domains of interest. The Centre works on cutting edge technology platforms and systems to build robust and innovative digital solutions for various domains like Agriculture, Animal Husbandry, Skilling, Education, Industry and governance, etc.

CDIPD specialize in designing comprehensive programmes to enhance the skills and capabilities of individuals and organizations in the digital space. It offers a diverse range of learning opportunities for the professionals looking to stay ahead in the ever-evolving digital domain and to organizations aiming to upskill their workforce. From hands-on workshops to in-depth training sessions, a spectrum of digital disciplines, including technology adoption, innovation methodologies, and industry-specific skillsets are covered. The goal is to empower individuals and teams to navigate the complexities of the digital era with confidence and competence.

11.5 K DISC

Kerala Development and Innovation Strategic Council (K-DISC) is the strategic think-tank advisory body of the Government of Kerala, formulating plans reflecting the latest in technology, product and process innovations for the development of the State. It seeks to address the challenges of providing holistic health care, employment, social security, infrastructure, food and nutrition security with a pro-poor bias, gender justice and inclusion of outliers, within severe fiscal constraints.

K DISC can give technical support to promote innovation and start up by students in this context Incubation centres can be started in identified colleges in each district where more facilities are available and neighbouring colleges can be connected to the incubation centre. A cluster of four or five colleges can be suggested for this. Colleges falling in each cluster should submit programmes/project to the lead college where incubation centre is located and enable the students to indulge in innovation, start-up and research.

Source: https://kdisc.kerala.gov.in

Government of Kerala has established Kerala Knowledge Economy Mission (KKEM) under Kerala Development and Innovation Strategic Council (KDISC) to co-ordinate skilling activities in the State and skill 60 lakh people in high end job skills and provide employment to 20 lakhs in the high end job roles in the state in five years.

The report of the Working Group on Knowledge Economy, Employment and Skills prepared by the Kerala State Planning Board for the 14th FYP is given in Annexure-7.

References

Aditi Jha 2024 Preparing India's workforce for an AI future, Observer Research Foundation, August 29, 2024

Anjali Raja 2024 Understanding the nature and scope of AI skilling programmes in India, https://indiaai.gov.in>article>understanding-the-nature

Agmatix The Power of Digital Agriculture, https://www.agmatix.com>blog>the-power-of-digital

Arthur G Green et al 2021 A scoping review of the digital agricultural revolution and ecosystem services: implications for Canadian policy and research agenda, FACETS, Vol 6, No.1, Dec 2021

ATRIA Agriculture 4.0- New Technologies in Agriculture, https://atriainnovation.com>blog>agriculture-4-0-new-technologies

Deloitte 2022 The Skills based organization: A new operating model for work and the workforce

Deloitte 2024 AI Skills: Interventions and Programmes needed, https://www.2.deloitte.com>analytics>

Higgins, V., Bryant, M, Howell, A and Battersby, J 2017 Ordering Adoption: Materiality, Knowledge and farmer engagement with Precision agriculture technologies, Journal of Rural Studies, 55, Oct: 193-202

LinkedIn 2022 Skills Advantage Report: Close skills gap at scale, https://learning.linkedin.com>future-of-skill-building

MSDE Skill India Digital Hub (SIDH), https://www.skillindiadigital.gov.in

World Economic Forum 2023 How AI is helping to identify skill gaps and future jobs, https://www.weforum.org

CHAPTER 12

GREEN SKILLS

Green skills are the knowledge, abilities, values and attitudes needed to live in, develop and support a sustainable and resource-efficient society. The need to transition towards more environmentally sustainable modes of production and consumption has become imperative, for developed as well as for developing countries, in the context of Sustainable Development (SD) goals. S D Goal 9 includes the target of upgrading infrastructure and retrofitting industries to make them sustainable, with increased resource-use efficiency and greater adoption of clean and environmentally sound technologies and industrial processes by 2030. S D Goal 12 includes the target of achieving the sustainable management and efficient use of natural resources by 2030.

The transition to a low-carbon, resource-efficient economy requires systemic changes that will result not only in new products and services but also changes in production processes and business models.

12.1. Green Skills explained

According to the ILO (2016), green jobs are decent jobs that contribute to preserve or restore the environment, be they in traditional sectors such as manufacturing and construction, or in new, emerging green sectors such as renewable energy and energy efficiency.

Green jobs are a priority for India, since the country is working towards its ambitious climate commitments to decarbonize the economy. Being one of the fastest growing major economies in the world, the third highest greenhouse gas emitter, and the fifth most vulnerable country to climate change impacts, India is moving towards green growth to decouple emissions from economic prosperity while also creating millions of decent jobs.

Green jobs discussions in India began in 2009, when the Ministry of Labour and Employment established the "Multi-stakeholder Task Force on Green Jobs and Climate Change" with ILO's support, to address employment and labour market dimensions of environment related policies and strategies.

12. 2. Skilling for Green Jobs

Skill Council for Green Jobs (SCGJ) is one of the most recently launched initiatives of the Government of India aligned to the National Skill Development Mission. It is promoted by the Ministry of New and Renewable Energy (MNRE) and Confederation of Indian Industry (CII). The creation of the SCGJ was approved by National Skill Qualifications Committee. Skilling in green sectors is led by SCGJ, although other ministries such as Ministry of Environment & Forests and Ministry of renewable energy are also implementing green skilling programmes in solar, wind, biogas, and forestry.

Integrating green skill requirements of industry into higher education and vocational training is critical to realizing India's future workforce requirements. Multiple government policies and initiatives across various sectors/ministries are pushing forth India's green agenda. In 2020, the Ministry of Environment, Forest and Climate Change (MOEFCC) constituted a high-level inter-ministerial Apex Committee for Implementation of Paris Agreement (AIPA) to generate a co-ordinated response on climate change matters that ensures India is on track towards meeting its obligations under the Paris Agreement including its NDCs, and to operate as a National authority to regulate carbon markets in India, issue guidelines on carbon pricing, market mechanism, and other similar instruments (MOEFCC, 2020).

In July 2022, India was part of the Labour and Employment Ministers Meeting (LEMM) declaration made by BRICS countries, of which one of the goals was to understand more deeply and promote green jobs and a Just Transition (ILO, 2022). Ministry of Labour and Employment has committed to better defining and measuring green jobs, strengthening green jobs and Just Transition research, adopting measures for human resource development, and classifying and creating standards for green jobs.

12.1 LEMM Declaration

The LEMM Declaration, issued at the conclusion of the meeting of Labour and Employment Ministers held in Sochi, Russia, on 9 and 10 September, 2024, emphasized four priority areas: enhancing skills development, protecting workers on digital platforms, promoting social security, and ensuring safe and healthy working environments, particularly for women.

"Recognizing the imperative need to adapt to the evolving labour market challenges," the BRICS member countries "commit to developing comprehensive strategies for the promotion of life-long learning, vocational guidance, continuous professional education and vocational skills training, enhancing modernization of employment services, ensuring safe and healthy working environments, and bolstering robust social support mechanisms," says the Declaration.

Source: https://g7g20.documents.org>database>document

In terms of labour market policy, awareness on upcoming green jobs is limited among trade unions. The Government has taken steps towards enhancing employability and training in green sectors by

i. working towards creating national level portals for wider dissemination of green jobs information such as SCGJ's Rozgar portal and Labour ministry's National Career Service portal,

ii. designing policies and incentives for on-the-job trainings for green skilling, particularly in the renewable energy and EV sector, and

iii. linking environmental conservation efforts with employment guarantee programmes. Occupational safety and health (OSH) and social protection policies are in place to some extent, but are missing specific strategies for regions affected by climate change and green transition.

12.3. Increasing green enterprise support

There is a need for separate and focused support for budding green enterprises in India through financial and mentorship assistance. Providing micro, small and medium enterprises (MSMEs) with trainings and financial incentives to green their businesses, creating green public procurement targets, and implementing policies to improve climate resilience of MSMEs in vulnerable geographies will increase scope for green jobs. Designing and supporting green entrepreneurship and apprenticeship programmes, by employers' organizations, will help in creating a pool of skilled green professionals. For industries, incentives should be designed to adopt greener practices.

12.4. Aligning skilling policies with climate commitments

A strategic roadmap is required to create a large pool of skilled workforce that meets India's NDCs and green targets. To scale up training programmes, the Skills Council for Green Jobs aim at (1) stronger industry collaborations in training and recruitment, (2) strengthening infrastructure and capacities of training partner network, and (3) obtaining green skilling requirements at a geography level to conduct localized trainings.

India is facing acute environmental challenges, including climate variability, poor air quality, over-exploitation of groundwater, water scarcity, increasing inland and coastal salinity, land degradation, and increased intensity of climate extreme events. India has made significant efforts to tackle these challenges through several policies and programmes, including, the National Policy on Biofuels, the National Clean Environment Fund, the Smart Cities Mission, the Green Hydrogen Policy, and concerted efforts to achieve its Nationally Determined Contributions (NDCs) as well as Sustainable Development Goals.

The Department of Science and Technology (DST) has been funding research and development (R&D) efforts since 2009 through its Clean Energy Research Initiative (CERI) to accelerate the pace of clean innovations. DST has been leading

Mission Innovation in Clean Energy which is part of the global Mission Innovation introduced at CoP21 in 2015.

The Ministry of New and Renewable Energy has funded R&D since 2017. As of November 2023, the Ministry of New and Renewable Energy has approved 50 solar parks. At CoP26 held in 2021, India committed to an ambitious five-part "Panchamrit" pledge, which included generating half of all energy requirements from renewables.

In 2019, MOEFCC released the draft National Resource Efficiency Policy (NREP) which aims to make industries resource efficient and also remediate pollution of air, water and land. It outlines regulatory and market-based instruments that can help optimize resource use at a sectoral level.

The Ministry of Skill Development and Entrepreneurship and the Ministry of MSME are key government stakeholders for enterprise policy. However, neither ministry has defined green enterprises in their policy documents. Though the Government support for green enterprise development is not strategic, multiple initiatives directly and indirectly promote greening in enterprises.

The Green Skills Academy affiliated to the National Skill Development Council offers entrepreneurship training programmes in green sectors such as solar energy, waste management and agriculture. The entrepreneurship arm of the Ministry of Skill Development and Entrepreneurship works extensively on building capacity of aspiring entrepreneurs and providing them with the necessary resources and government/industry connections. While they do not have specific support for green enterprises, they strongly encourage green startups. Their target groups include women, unemployed youth, and persons from the LGBTQ+ community.

12.5. Green entrepreneurship initiatives

UN agencies have worked on assistance with green entrepreneurship, as below:

- The ILO's Start and Improve Your Business programme (SIYB) trains aspiring entrepreneurs in business with a focus on youth and women from rural areas. The programme has launched a Green Business booklet to help entrepreneurs come up with a green business idea and existing entrepreneurs to green their businesses.
- UNICEF India has been working on educating and mobilizing India's youth about green careers, connecting youth to professionals in the field, and building their capacities to ably contribute to the environmental sector
- UNEP India has worked on highlighting the voices and journeys of young green entrepreneurs, and has created an Eco-Entrepreneurship Guide to assist greening of SMEs (not specific to, but is applicable to India)
- UNDP consistently engages with young green entrepreneurs, connecting them with aspiring entrepreneurs. UNDP is currently working on a green recovery project aiming to train rural entrepreneurs in the EV sector.

Outside the Government and the UN agencies, start-up accelerators and incubators such as the Climate Collective have a sole focus on climate/green solutions, with a strong focus on supporting women green entrepreneurs. Climate Collective is a dynamic team of nearly 40 driven individuals, who daily build platforms that support climate tech entrepreneurs in the early perilous stages of the startup journey.

The Green General Skill Index identifies four groups of work tasks especially important for green occupations:

- Engineering and technical skills: hard skills encompassing competences related with the design, construction and assessment of technology usually mastered by engineers and technicians. This know-how is needed for eco-buildings, renewable energy design and energy-saving research and development (R&D) projects.
- Science skills: competences stemming from bodies of knowledge broad in scope and essential to innovation activities in physical and biological sciences. These skills are in high demand in each stage of value chain and in the utility sector, which provides basic amenities such as water, sewage services and electricity.
- Operation management skills: know-how related to change in organizational structure required to support green activities and an integrated view of the firm through life-cycle management, lean production and cooperation with external actors, including customers. Such skills are important for sales engineers, climate change analysts, sustainability specialists, agricultural officials and transportation planners.
- Monitoring skills: technical and legal aspects of business activities fundamentally different from the remit of engineering or of science. They refer to skills required to assess the observance of technical criteria and legal standards. Examples are environmental compliance inspectors, nuclear monitoring technicians, emergency management directors and legal assistants.

In addition to these skills, a range of soft skills are also considered increasingly important, not only for green skills, but generally for 'skills of the future', including also those necessary for the Fourth Industrial Revolution. In particular, skills related to design thinking, creativity, adaptability, resilience, and even empathy, are regarded as critical.

India is a signatory to several international conventions on biodiversity and climate change and the goals set therein. However, governments alone cannot achieve them. While environment protection and conservation is an important task of the government, the Indian Constitution also clearly imposes duty on every citizen to protect environment. Peoples' involvement is therefore, essential. Consequently, the prerequisite is to empower people with skills on environment protection and conservation.

The National Education Policy 2020, in revamping the current education system to be more holistic and multidisciplinary, suggests integration of greening into school and university curriculum. At school level, existing environmental curriculum will be updated with contemporary topics such as climate change, waste management, forest/wildlife conservation, and sustainable living. At the higher education level, focus will be on enhancing employability through skilling in environment and sustainable living for undergraduate students. The policy prioritizes courses on disruptive technologies such as renewable energy, water conservation, sustainable farming, and environmental preservation.

Realizing the need for developing the green skills in different areas, the MoEFCC in June, 2017 launched, on a pilot basis, the Green Skill Development Programme (GSDP) under the ongoing Environmental Information System (ENVIS) Scheme to skill youth in environment, forest and wildlife sectors and enabling them to be gainfully employed or self-employed. The programme was conducted at ten locations, spread over nine bio-geographic regions of the country. The pilot programme offered a Basic Course and an Advanced Course of three months' duration each, to skill the youth as Biodiversity Conservationists and Para-taxonomists, respectively. The Botanical Survey of India (BSI) and Zoological Survey of India (ZSI), at the headquarters and their regional offices, were the nodal centres for the pilot programme.

With the success of the pilot run, the programme was expanded to an all India level with the launch of the full-fledged programme as well as the Mobile app on GSDP in May, 2018. The app (gsdp-envis) contains all basic information about the training programmes being conducted under GSDP.

Currently 44 GSDP courses are offered at 87 locations across the country. With duration ranging from 80 hours to 550 hours, they cover diverse fields. These include, Pollution Monitoring (air/water/soil); Emission Inventory; Sewage Treatment Plant (STP)/Effluent Treatment Plant (ETP)/Common Effluent Treatment Plant (CETP) operation & maintenance; Waste Management; Environmental Impact Assessment; Forest Fire Management; Water Budgeting & Auditing; Conservation Of River Dolphins; Wildlife Management; Para Taxonomy Including Peoples Bio-Diversity Register (PBR); Mangroves Conservation; Bamboo Management & Livelihood Generation; Valuation Of Ecosystem Services; Water Budgeting & Auditing, etc.

All skilling programmes are being aligned with the National Skills Qualifications Framework (NSQF), requirement of the National Skill Development Agency (NSDA) and Ministry of Skill Development and Entrepreneurship (MSDE).

The UK is in a race to decarbonise its operations by 2050. In light of the UN's SDGs, increasing Environmental, Social and Governance (ESG) regulations and penalties, and an incoming wave of consumers and job candidates that prioritise sustainability, environmental action has become a non-negotiable for modern businesses.

However, demand for workers with sustainability expertise is outpacing the supply of talent. This shortage has been termed 'the green skills gap'. The green skills gap is the shortage of workers needed to fill incoming jobs created by the green transition - a shift away from reliance on fossil fuels and towards a sustainable future.

According to the World Economic Forum, the global number of green roles has grown 8 per cent per year for the last five years. However, there are not enough employees and candidates with the right skills needed to fulfil these crucial roles.

During July 2024, the British Council, in partnership with HSBC India, launched the *Climate Skills: Seeds for Transition India* project in New Delhi. This global initiative aims to nurture essential climate skills among India's youth, creating over 2,000 youth leaders from marginalised communities and training 300 educators in India over the next two years. The project will also engage young people aged 18-30 years in Brazil, Indonesia, Mexico, and Vietnam. The programme is a part of the British Council's long-term international effort to establish a global climate skills approach. It seeks to empower youth, inform decision-makers about alternative climate education strategies, and address climate challenges at grassroots, national, and global levels. By partnering with governments, communities, and institutions, the initiative aims to create tangible solutions to better understand the impact of climate change and equip the next generation with necessary skills needed in the wake of climate change.

Climate Skills equips young people around the world with the fundamental skills needed to be part of the transition from high carbon to low carbon economies. Through a collaborative effort between the British Council and HSBC India, the programme trains local community organisations to better support young people who are disproportionately affected by climate change.

The participants come from communities that encounter major systemic challenges in contributing to the growth of their country's green economy. The programme will delve into climate literacy, helping to foster a deeper understanding of shared environmental challenges and solutions, and nurture core skills and leadership abilities, providing a strong foundation for personal and professional growth.

12.2 PAGE

The Partnership for Action on Green Economy (PAGE) was launched in 2013 as a response to the call at Rio+20 to support those countries wishing to embark on greener and more inclusive growth trajectories. PAGE brings together five UN agencies: UN Environment Programme (UNEP), International Labour Organization (ILO), UN Development Programme (UNDP), UN Industrial Development Organization (UNIDO), and UN Institute for Training and Research (UNITAR) whose mandates, expertise and networks combined can offer integrated and holistic support to countries on inclusive green economy, ensuring coherence and avoiding duplication. ILO's activities within PAGE focus on green jobs and ensuring a Just Transition, one that manages workplace changes as a result of the impacts of climate change and climate action. India joined PAGE in 2018 to catalyse action and enhance cooperation on existing initiatives on national and subnational levels.

Source: https://www.un-page.org

India, known for its rich and diverse ecosystems, is facing severe environmental challenges due to rapid urbanisation and development. The brunt of these environmental challenges is felt by the youth, who have to deal with not only the immediate effects but also the long-term consequences. Thus, for India's younger generation, building resilience against these crises is vital, as they will inherit and shape the planet's future. In this context, green skills have emerged as an essential tool in mitigating and adapting to the rapidly changing climate. These skills encompass technical knowledge, practical competencies, and mindsets required to implement sustainable practices, environmentally friendly technologies, and resource-efficient solutions across various sectors.

The Network for Certification and Conservation of Forests (NCCF) is a registered society of experts, professionals, and stakeholders of natural resources, engaged in the development of India-specific and globally benchmarked sustainability standards, as its core working area, apart from working for policy advocacy, training and capacity building through multi-stakeholder engagement. NCCF has developed the first ever Indian Certification Standard for Forest Management endorsed by the leading international certification organization, the PEFC, giving it global recognition.

NCCF does the capacity building of the stakeholders through National, State, regional level workshops, Training, and Skill development to sensitize stakeholders about Certification and conservation of Natural resources. NCCF is conducting MoEFCC's Green Skill Development Programme for a two-month Certificate Course in VRIKSH –Timber Legality Assessment Programme.

NCCF has developed a Carbon Registry-India (CR-I) to provide a platform for carbon tracking and trading for verified net Green House Gases emissions reductions and removals enhancement with Sustainable Development Contributions. CR-I will also facilitate micro-scale projects and has the potential to provide a compatible platform for small landholder farmers to avail carbon credits and get linked to the carbon markets. The Registry will help in climate change mitigation and adaptation and aims to diligently follow up with India's NDCs.

12.6. Green skill development programme in India

As the world faces environmental challenges such as climate change, natural resource depletion and pollution, sustainable development is becoming increasingly important. India has made significant progress in sustainable development in recent years by promoting renewable energy, reducing greenhouse gas emissions, improving access to clean water and sanitation. One of the key initiatives by India towards sustainability is the Green Skill Development Programme (GSDP).

MoEFCC launched GSDP in 2017 which is implemented through the NSDC. The programme aims to create a pool of skilled and certified workforce in the environment and forest sector, including biodiversity conservation, pollution management, and renewable energy. It focuses on providing training and certification in various "green skills" to the youth, particularly from rural and underprivileged communities for sustainable development of the country.

Green skills, also known as sustainable skills or environmental skills, refer to the knowledge, abilities, and competencies required for individuals to contribute to a more sustainable and environmentally friendly society and economy.

These skills encompass a range of technical and soft skills that enable individuals to understand, promote, and implement environmentally responsible practices across various industries and professions.

Green skills are essential for addressing environmental challenges, such as climate change, resource conservation, pollution reduction, and sustainable resource management. They play a crucial role in promoting a more eco-conscious and sustainable future.

Types of green skills

Green skills include a wide range of competences that can be divided into two categories: technical skills and soft skills. Both are required for individuals to contribute to environmental sustainability and responsibility.

Technical Skills:

Technical green skills are industry-specific competencies that enable people to engage directly with technology, processes, and practices that reduce environmental effects. These abilities are especially beneficial in industries that have a direct impact on the environment, such as energy, construction, waste management, and agriculture. Here are some examples of technical green skills:

- Renewable Energy Technologies: Knowledge of how to install, maintain, and operate renewable energy sources such as solar panels, wind turbines, and hydropower systems.
- Energy Efficiency: Knowledge of how to optimize energy usage in buildings, industrial operations, and transportation in order to reduce energy waste and greenhouse gas emissions.
- Sustainable Agriculture: Understanding of eco-friendly farming practices that lessen agriculture's environmental impact, such as organic farming and crop rotation.
- Waste Management: Recycling, composting, and waste reduction skills to reduce landfill waste and encourage resource conservation.

- Green development and Design: Knowledge of planning and developing sustainable structures that maximize energy efficiency, employ environmentally friendly materials, and use green technologies.

Soft green skills

Non-technical competencies that enable individuals to traverse the complicated landscape of sustainability, communicate effectively, and collaborate to execute environmental solutions are referred to as soft green skills. These abilities are useful in any industry and are critical for promoting environmental initiatives.

- Communication: The ability to communicate complex environmental concepts and efforts to a wide range of audiences, including stakeholders, colleagues, and the public.
- Teamwork and Collaboration: Effectively addressing environmental concerns and implementing sustainability projects with varied teams and stakeholders.
- Problem Solving: The ability to recognize environmental difficulties, analyze data, and devise innovative solutions to address sustainability issues.
- Adaptability: the ability to keep up with changing environmental rules, emerging green technologies, and evolving best practices.
- Leadership: guiding and encouraging others to embrace sustainable practices, as well as advocating for environmental responsibility within organizations and communities.

Technical and soft green skills are both required to build a sustainable and environmentally responsible future. Individuals with a combination of these talents can address environmental concerns, decrease their environmental imprint, and contribute to a more sustainable and environmentally conscious society.

The GSDP has identified 80 different green skills that are in demand in India. These skills are categorized into various sectors like biodiversity conservation, environmental management and pollution control, renewable energy, energy efficiency, waste management, sustainable agriculture, climate change mitigation and adaptation, sustainable water management, eco-tourism, sustainable building designs and sustainable construction

The GSDP is designed to cater to the diverse needs of the industry and the environment sector, and is aligned with the National Skills Qualification Framework (NSQF) to ensure standardization and recognition of skills.

The GSDP has a wide scope and covers various regions and communities across India. It has partnered with several institutions, both government and non-governmental, to implement the programme. These institutions include the Indian Institute of Technology (IIT), Indian Institute of Forest Management (IIFM), Wildlife Institute of India (WII), the Centre for Environment Education, and the National Institute of Solar Energy, among others. The program has also collaborated with state governments particularly the forest departments and pollution control boards to ensure that it reaches all regions of the country.

Benefits of GSDP

The GSDP has several benefits, both for individuals and for the country as a whole. For individuals, it provides an opportunity to gain new skills and knowledge in a growing sector. It also enhances their employability and opens up new job opportunities. For the country, it creates a skilled workforce that can contribute to sustainable development. It also promotes the conservation and management of the environment.

The GSDP has helped contribute to the growth of green economy in India. According to a report by the NSDC, the programme has trained over two lakh candidates in various green skills, and has placed more than 60,000 candidates in jobs in the environment sector. The programme has also created self-employment opportunities for many of the trainees, who have started their own ventures in areas such as organic farming, eco-tourism, and renewable energy.

Quest Alliance has introduced a new green curriculum which aims to build green mindsets in young learners. The Green Innovation project at Government Industrial Training Institute, Halol, Gujarat is an initiative to drive green mindsets and inculcate an education in young people that will enable them to pursue green jobs.

The modules called 'Green Futures' are spread across seven sections, as below:

- Introduction Green Mindset
- Look into the Future – Agriculture and Industries
- The Present and the Future
- Waste Management
- Green Practices & Green Jobs
- A Look into Future Trends
- Sustainable Resources

A successful transition towards a greener world will depend on the development of green skills in the population. While green skills are relevant for people of all ages, they have greater importance for younger people as they can contribute to the green transition for a longer period. Youths play a crucial role in shaping the future, and their preparedness to embrace green skills is vital for creating a sustainable world.

Green skills in the agricultural sector focus on sustainable farming practices, organic agriculture, permaculture, agroforestry, and efficient resource management, etc. Professionals with these skills help promote sustainable food production, biodiversity conservation, and responsible land use.

The global push towards sustainability and the reduction of carbon footprints is influencing job markets. The 'green economy' encompasses renewable energy, sustainable transportation, energy efficiency, waste management and green construction. This shift towards sustainability is driven by the need to combat climate change, conserve resources and promote environmental responsibility.

The challenges of climate change require multilateral solutions, including creating resilience funds, de-risking platforms, promoting circular economies, joint technological development, and enhancing green energy security and transition partnerships. However, to cope with the disproportionate impacts of climate change, setting up a "flexibility mission" is important for these countries. This would provide them with the means to adapt and innovate in response to environmental shifts

While Indian agriculture is adversely impacted by the vicissitudes of climate change, the sector also is a significant contributor to greenhouse gas (GHG) emissions. As per the Third Biennial Update Report submitted by the Government of India in the early 2021 to the United Nations Framework Convention on Climate Change (UNFCCC), the agriculture sector contributes 14 per cent of the total GHG emissions.

The National Mission on Sustainable Agriculture, as part of National Action Plan on Climate Change for more than a decade now, has focused to make Indian agriculture sustainable, considering likely risks arising from climate variability.

The list of strategies/pathways in the policy agenda to make Indian agriculture green and carbon neutral are:

1. Diversification to crops like pulses, oilseeds, vegetables and fruits, adapted to specific agro-ecologies need to be planned, and implemented by the states with suitable incentives to farmers during the changeover. However, diversifying to new portfolio of crops will require strengthening of existing value chains.
2. Conservation agriculture offers solutions to many soil health problems with good agronomy and soil management such as no-till farming, crop rotation, in-situ crop harvest residue management / mulching, zero-till planters, etc.
3. Water use efficiency- "More crop-per drop" has been the mantra of current public policies around irrigation water. Suitable policies with incentives mechanism could lead more farmers to adopt technologies that aim to "irrigate the crop and not the land".
4. Setting up of solar power plants on farmlands, wherever possible, and solarising existing grid-connected pumps, could earn additional income to farmers, besides making them net energy producers.
5. **Digital agriculture:** Increasing use of mobile telephones (and smartphones) in rural India offers a unique opportunity to leverage information symmetry and connectivity to the advantage of farmers.

6. Make farming carbon neutral: Carbon-neutral farming refers to agricultural practices that aim to balance the amount of carbon dioxide emitted during farming operations with the amount of carbon captured and stored, particularly in soil, trees, and plants. This balance is achieved through techniques such as sustainable soil management, agroforestry, and renewable energy use, which not only reduce emissions but also sequester carbon in ecosystems.

The agriculture sector plays a critical role in addressing climate change, as it accounts for significant greenhouse gas emissions, primarily through deforestation, livestock methane, and fertilizer-related nitrous oxide. Embracing carbon-neutral practices empowers farmers to fight climate change, improve soil health, enhance resilient farming systems, and secure food for the future.

References

Centre for Policy Research, Initiative on Climate, Energy and Environment, https://cprindia.org>Researchinitiatives

ICRISAT Digital Agriculture- Tools and Technologies, https://www.icrisat.org>research>tools-and-technologies

IEA 2023 India's Skill Council for Green Jobs https://iea.org>policies>17819-india-skill

ILO 2011 The Green Jobs Programme of the ILO https://sustainabledevelopment.un.org>ilo

ILO 2016 What is a green job? https://www.ilo.org/global/topics/green-jobs/news/wcms220248/lang-en/index.hn

ILO 2018 Skills for green jobs in India

Ministry of Power 2022 Green Hydrogen Policy

Ministry of FF&CC The National Mission for Green India, https://moef.gov.in>about-the-mission

NITI Aayog 2019 Strategy for New India@75

PAGE 2023 Assessment of India's Green jobs and Just Transition Policy Readiness, ILO

PAGE 2023 What is PAGE doing to promote a Just Transition? https://www.un-page.org>news>what-is-page-doing

Skill Council for Green Jobs 2022 Green Jobs Handbook

CHAPTER 13

SKILLS FOR THE FUTURE

This chapter is presented in two sections- Section A presents a general overview of the future work scenario, where soft skills need to be emphasised and Section B presents the scenario of adaptation to climate change and disaster management in the agriculture sector.

Section A

The future of work is expected to be a combination of Force, Space and Culture. Force comprises of collaboration between human beings and technology. Automation of work will decide the velocity of force. Space is referred as connectivity, digital assistance, content and collaboration. It relates to connected and secure work environment. Culture refers to the empowered workers aligned to new digital skills. The focus is on skill acquisition, development and retention.

13.1. Comparison of present and future work situation

The table below makes a comparison of the present and future work situation

Present	Future
Employee engagement	Employee experience
Hard skills	Soft skills
Definite hours and location	Flexibility
Manage	Coach and Mentor
Humans work like robots	Humans work with robots and AI
Process centric tasks	Strategic and creative tasks
Hierarchy	Flat structures
Static workforce	Dynamic workforce
Company functions like a factory	Company functions like a laboratory
Move for job	Move for job and life

Source: www.worldskillcouncil.org

The challenges that workforce and leaders face today are more significant and complex than they have been earlier. Global crises such as climate change and disasters demand action. Accelerating the digital journey of the workplace carries tremendous execution risks. Burnout has become a epidemic, now recognised by WHO as an official disease. The anxiety and weariness of the workers affect the job performance.

Organisations have started to focus on developing human skills a lot more than they have done in the past. Human skills such as curiosity, insight, empathy and resilience come to the forefront due to automation and other factors. If the workforce is skilled in this domain, they will be more agile in their approach to delivering value to customers.

According to a report by Price waterhouse Coopers (PwC), there are four worlds of work in the future- Red, Blue, Red and Yellow.

13.2. Four Worlds of Work

The table below presents the key elements of the four worlds of work.

World	Key Element	Description
Red	Innovation rules (perfect incubator for innovations)	- Use of technology and IP rather than human effort - Contract negotiations are key - Skill enhancement through open source learning - Specialism is highly prized - Small number of pivotal people with outstanding management skills command high rewards
Blue	Corporate is King (Capitalism reigns supreme)	- Talent is bought when needed - Talent search for exceptional talent - Human efforts, automation, analytics and innovation are key for performance - Success depends on productive workforce as companies compete for the best talent - Era of super workers - Employees take active role in their career development, honing their skills
Green	Companies care (Corporate responsibility is a business imperative)	- Employees enjoy family-friendly flexi hours - Trust and ethics play the central role - Well-being support like sleep clinics, digital dieting, etc adopted - Travel minimized, virtual social networks among the organisation and client base - Take part in socially useful projects

Yellow	Humans come first (New worker guilds thrive)	- Seek out greater meaning and relevance in what is done - Employees feel loyalty to people with the same skills or cause - Like-minded workers gravitate towards each other - Digital platforms create mobility

Source: PwC report "Workforce for the future- The competing forces shaping 2030

13.3. Skills for Resilience

Resilience means being able to cope with tough events. When you have resilience, you harness the inner strength that helps you rebound from a setback or challenge. When something bad happens, you still feel anger, grief and pain. But you are able to keep going, both physically and psychologically.

Developing resilience can help you cope adaptively and bounce back after changes, challenges, setbacks, disappointments, and failures. Developing skills of resilience can help you face challenges and difficulties in life, which can help you feel better and cope better.

There are many ways to increase resilience. Some of those include having a good support system, maintaining positive relationships, having a good self-image and having a positive attitude. Developing resiliency can help you maintain caring relationships with others and help you maintain a positive and easy going disposition. It can also help you develop good coping skills and improve cognitive thinking skills.

13.4. VUCA Environment

The leaders face numerous challenges in the unpredictable and rapidly changing world and business environment. One of the most prominent frameworks used to describe these challenges is VUCA leadership.

VUCA stands for Volatility, Uncertainty, Complexity and Ambiguity, and it captures the key elements that leaders must navigate to succeed.

Volatility - Sudden and unpredictable changes

Uncertainty - Lack of predictability or clarity about the future

Complexity - Intricate and interconnected nature of situations

Ambiguity - Multiple possible interpretations or meanings

Volatility refers to the rapid and unpredictable changes that can occur in the business world. This could include sudden market shifts, economic instability, technological advancements, or other unforeseen events that can significantly impact an organization's operations, strategies, and overall stability.

A startup company connects with a top app development company to build an effective and useful app for its startup business. Suddenly, a major innovation transforms the procedure of development of apps. This abrupt change creates volatility. The leaders of the startup must swiftly adapt their strategies to stay competitive in the new landscape.

In the context of VUCA world, **uncertainty** refers to the lack of predictability or clarity about future events or outcomes. It implies that leaders may not have access to all the information they need to make confident decisions, and there is a level of ambiguity surrounding potential outcomes.

A high profile global company manufactures products using spare parts from different countries. A political tension in a region affects the supply chain. The leaders of the company won't be sure how it will impact the business. This makes it harder to plan ahead, and hence they have to be ready to adjust their business strategies as things unfold.

Complexity involves dealing with many interconnected parts, systems and relationships. These elements can be challenging to understand or manage because they often have dependencies on one another. With the web of interrelated factors, it is essential to have a clear understanding of how different components interact. Complexity can arise from various sources, such as global supply chains, diverse stakeholder interests, or rapidly evolving technologies.

For example, a big company with presence all over the world has to co-ordinate research, production, logistics, and more such activities across different regions. The company also faces various government regulations, cultures, and consumer preferences. Managing all these aspects is an example of dealing with complexity in modern world of business. Leaders need to view the big picture while understanding the details to navigate such situations effectively.

Ambiguity refers to situations where information is unclear, vague, or open to interpretation. It means that there may be multiple possible meanings or ways to interpret a given situation or piece of information. It can arise from various factors, such as conflicting information, changing market dynamics, or evolving customer needs.

A company is considering entering a new market, but the economic conditions are not clear. Some signs suggest growth, while others point to risks. Leaders must carefully weigh the available information, consult experts, and conduct thorough research. They may need to develop contingency plans for different outcomes. This showcases their ability to make decisions in ambiguous situations and manage risks in uncertain environments.

VUCA leadership approach requires leaders to be adaptable, flexible, and resilient in the face of constant uncertainty. VUCA leaders must be able to navigate challenges and make informed decisions despite limited information and shifting circumstances.

The VUCA framework provides a lens through which leaders can navigate these challenges and manage a rapid and unpredictable change. By understanding the four elements of VUCA and their impact on leadership, leaders can develop the necessary skills and mindset to thrive in this dynamic environment.

Skills for the VUCA environment

To thrive in a constantly changing VUCA world, leaders need to develop specific skills that enable them to prevent the negative consequences of the dynamic nature of business environment. The skills are:

1. Agility and Adaptability

An agile leader is one who can adapt quickly to changing circumstances and seize opportunities as they arise. This requires a willingness to experiment, take calculated risks, and learn from failures.

Agility in leadership goes beyond just being flexible. It involves having a growth mindset and being open to new ideas and perspectives. VUCA leaders with agility understand that the business landscape is constantly evolving, and they proactively seek out innovative solutions to stay ahead of the curve.

Furthermore, agile leaders are not afraid to challenge the status quo. They encourage their teams to think creatively and push boundaries, fostering a culture of continuous improvement.

2. Strategic Thinking and Decision-Making

In a VUCA world, leaders must possess strong strategic thinking skills, which involve anticipating challenges, identifying trends, and considering both external factors and internal capabilities. This requires a balance of analysis and intuition.

Effective strategic thinking entails weighing various perspectives and analysing data for informed decision-making. VUCA leaders must excel at assessing risks and opportunities, enabling them to make sound judgments in complex and uncertain situations.

Strategic thinking extends beyond short-term goals. VUCA leaders need to prioritize long-term vision, aligning decisions with the organization's overall strategy. They must proactively identify emerging trends and adjust plans to ensure their teams are prepared for future challenges.

3. Resilience

Resilience encompasses the ability to bounce back from setbacks and maintain composure under pressure. Resilient leaders continue to lead effectively even in challenging situations.

Resilience is not just about enduring difficulties; it involves actively finding ways to adapt and thrive. It requires a blend of mental and emotional fortitude, as well as the capacity to learn from experiences and use them as opportunities for growth.

A valuable source of information is Bob Johansen's book "Leaders Make the Future". His ideas can be referred in formulating advice on good strategies to face the VUCA world. (Annexure-6)

13.5 Key Skills for Thriving in a VUCA World

Adaptability

The ability to quickly adjust to new conditions, environments, and challenges is crucial. This includes being open to change, being comfortable with ambiguity, and being able to pivot when necessary.

Resilience

Developing resilience helps individuals bounce back from setbacks and failures. It involves maintaining a positive attitude, managing stress effectively, and persisting in adversity.

Critical Thinking

This skill involves analysing information objectively, evaluating different perspectives, and making informed decisions. In a VUCA world, discerning reliable information from noise is essential.

Creativity and Innovation

Thinking creatively and generating new ideas is vital for staying ahead in a rapidly changing world. This includes seeing problems from different angles and devising novel solutions.

Collaboration

Working effectively with diverse teams and across different cultures is the key. Collaboration fosters innovation, allows for the pooling of diverse talents, and helps navigate complex challenges.

Emotional Intelligence

Understanding and managing emotions, both your own and those of others, is crucial for effective leadership and teamwork. This includes empathy, self-awareness, and the ability to build strong relationships.

Digital Literacy

In today's VUCA world, being comfortable with technology is essential. This includes knowing how to use digital tools and understanding emerging technologies and their impact on industries and society.

Strategic Thinking

It is important to view the bigger picture, anticipate future trends, and formulate long-term plans. Strategic thinking helps make decisions that align with organisational goals and adapt to changing environments.

Continuous Learning

A commitment to lifelong learning is necessary to stay relevant in a VUCA world. This includes being open to new ideas, seeking growth opportunities, and adapting skills to evolving demands.

Decision-Making

It is critical to make effective decisions in uncertain situations. This involves weighing risks, gathering relevant information, and considering multiple perspectives before taking action.

Agility

It is crucial to move quickly and decisively in response to changing circumstances. This includes being agile in processes, structures, and mindset.

Visionary leadership becomes crucial when leading teams through uncertainty with a distinct sense of purpose and direction. Proficiency in communication is crucial for promoting comprehension and unity in the face of complexity, and emotional intelligence and empathy are necessary for establishing connections with people from different backgrounds in the quickly changing settings. The capacity to foster innovation and promote an ongoing learning culture enables leaders not only to endure, but also prosper in the face of volatile, uncertain, and complex environments, transforming adversity into favourable circumstances.

Global and Cultural Awareness

Understanding diverse cultures, global trends, and geopolitical dynamics is important for businesses and individuals operating in a VUCA world.

According to World Skill Council, problem solving, self-management, working with people, technology use and development are some of the skills among the top Skills of 2025. The emphasis is on the emerging skills requirement, as the very basic nature of work and meaning of workplace is in transition stage. The rules of the game have changed.

One outcome of the accelerated pace of technology's advancement is that it creates an environment of VUCA.

Technology advancements cause transformation. Transformation equates to change. But it is known that change either good or bad, drives strong emotions, which can also be positive or negative. Other skills, such as soft skills, are important to the success of technological transformations.

Soft skills (more recently referred to as "power skills" or "people skills") are needed to balance any technology transformation. With the rapid advancement of machine learning and artificial intelligence, many employees are concerned about their job security. There is no doubt that technology is transforming the way people work. Yet, power skills are still required. Learning and development leaders must recognize that increased power skills can make the technology transformation work for the organization, instead of against their workforce.

There are many power skills beneficial in a VUCA situation. Empathy allows leaders to ease employees into future workflow changes. Emotional intelligence creates better working relationships as employees adjust to their new normal. These types of skills help develop employees who are open to learning something new, seek more information and are ready to succeed.

A VUCA environment can:

- Destabilize people and make them anxious
- Sap their motivation
- Thwart their career moves
- Make constant reskilling and upskilling a necessity
- Require significant amounts of time and effort
- Increase the opportunities for people to make poor decisions
- Paralyze decision-making processes
- Jeopardize long-term projects, developments and innovations
- Overwhelm individuals and organizations
- Take a toll on internal culture

The traditional strategic planning process is no longer viable. Leaders need to create organizations that are agile, innovative, adaptable and people-centric. Today's dynamic, ever changing business environment needs a different style of leadership.

13.6. From VUCA to BANI

Over and above the dizzying succession of crises and emergencies, presently, there seems to be a shift in the way of thinking. For a long time, global governance issues were considered from the angle of the VUCA model. The shift from a VUCA to a BANI world reflects the evolution of an unstable environment increasingly fragile, anxious, non-linear and incomprehensible.

According to the futurist writer Jamais Cascio, it is 'difficult to see the big picture when everyone insists on colouring outside the lines'. Cascio writes about the intersection of emerging technologies, environmental dilemmas and cultural transformation. For him, as for others, the acronym VUCA is now obsolete. The reason is the emergence of new 'situations in which conditions are not only unstable', but also 'chaotic'. This modern reality of chaotic situation is best described by the new acronym BANI. The world is now:

- Fragile (B for brittle), with multiple disasters and regular disruptions, highlighting the precariousness of organisations.
- Anxious (A), with constant access to unpleasant information.
- Non-linear (N), where nothing is certain or can be assessed in advance. The links between cause and effect become confused.

- Incomprehensible (I), with events that are illogical, unpredictable and inexplicable.

Brittle reflects the illusion of strength. It refers to the myth that people tell themselves and each other to feel better and more secure. Resilience is the key to tackle with brittle environment. Anxious refers to the illusion of control.

We are living in a world loaded with too much information, threats and fake news, which results in anxiety. Emotional intelligence helps in better tackling and avoiding anxious environment.

Non-linear refers to the illusion of predictability. The connection between cause and effect is not clear. One should be prepared to move forward and back as the situation demands. Incomprehensible reflects the illusion of knowledge. It relates to people's experience that they do not understand what is going on. Individuals experience an incomprehensible situation when we struggle to find answers, but the answers are not convincing.

BANI scenario demands a unique set of skills to survive in a dynamically changing environment. They are:

- Creativity: need to be creative in solving problems with unique solutions
- Emotional intelligence: Understanding the emotions of self and others is a prerequisite for survival.
- Resilience: Moving forward encountering setbacks determine success
- Adaptability: Need to align goals, plans and strategies to stay in the competition
- Collaboration: 'I' should be replaced by 'We'. Shared goals and shared results will flourish
- Continuous learning: What to learn, how to learn and unlearn is a required skill
- Agility: Swiftness is needed. Moving quickly and adapting to changing environments will give a competitive advantage
- Systems thinking: Individuals to be geared up to systems thinking. Complex interactions between different stakeholders take place
- Critical thinking: Complex and ambiguous situations to make informal decisions
- Risk management: Anticipation and management of risk should be a skill of high order
- Communication: Communicating ideas, messages and values are important
- Transparency: To be maintained in communicating fear of challenges and problems

- Co-operation: Synergy is the energy. Esprit de corps should be the voice of every team member.
- Empathy: Listening and communicating emotionally will bind the organisation.

Section B

13.7. Adaptation of Farmers to the Unpredictability of Climate Change

There are two terms used in relation to climate change: Adaptation and Mitigation. Adaptation, in the simplest terms, refers to the actions that countries will need to take to respond to the impacts of climate change already happening, which are very much felt and experienced and at the same time, preparing for future impacts. It refers to changes in processes, practices and structures that can reduce the vulnerability of the farmers to climate change impacts, such as sea level rise or food insecurity.

Mitigation refers to efforts to reduce or prevent emission of greenhouse gases. Mitigation can mean using new technologies and renewable energies, making older equipment more energy efficient, or changing and adjusting management practices or consumer behaviour.

Climate change is a reality that farmers are grappling with daily. The farmers demonstrate a remarkable resilience and adaptability in the face of the uncertainty brought about by climate change. Some of the ways in which farmers adapt to climate change include:

- Diversification of crops and livestock: Diversification provides a buffer against crop failures. Farmers are experimenting with new crops and varieties better suited to the changing conditions.
- Precision Agriculture and Technologies: Technological advancements have played a crucial role in helping farmers adapt to climate change. With data-driven insights from weather forecasts and GPS-guided equipment, farmers can make informed decisions about the time of sowing and other cultivation practices to be followed in the field.
- Sustainable farming practices: Conservation Agriculture is the need of the hour. It is a farming system that can prevent losses of arable land while regenerating degraded lands. It promotes maintenance of a permanent soil cover, minimum soil disturbance, and diversification of plant species. It enhances biodiversity and natural biological processes above and below the ground surface, which contribute to increased water and nutrient use efficiency and to improved and sustained crop production. It opens increased options for integration of production sectors, such as

crop-livestock integration and the integration of trees and pastures into agricultural landscapes.

- Climate resilient crops: Farmers are adopting climate resilient crops at increasing rates to mitigate the impacts of climate change. The breeders are involved in developing climate resilient varieties of different crops that can withstand extreme weather conditions.
- Weather forecasts and Early warning signs: Recent developments in technology, specifically artificial intelligence (AI), have brought new attention to weather forecasting, for which the Numerical Weather Prediction (NWP) models has remained the primary process for many decades. AI and machine learning are now being applied to forecasting, not only using numerical models and atmospheric physics but incorporating historical data as well.
- Capacity building: National Innovations on Climate Resilient Agriculture (NICRA) is a network project of the ICAR, which aims to enhance resilience of Indian agriculture to climate change and climate variability through strategic research and technology dissemination. The activities on adaptation and mitigation covers crops, livestock, fisheries and natural resources. Capacity building of farmers is an important activity under NICRA.

The Government is implementing the National Action Plan on Climate Change (NAPCC) which provides a policy framework for climate action in the country. National Mission for Sustainable Agriculture (NMSA) is one of the important missions under the NAPCC to make Indian agriculture more resilient.

Indo German Global Academy for Agroecology Research and Learning (IGGAARL) established in Kadapa, AP aims at developing 'climate resilent villages' through capacity building of farmers (Annexure-9).

13.8. Resilience approach to disaster risk management in Agriculture

National Institute of Disaster Management (NIDM) has developed a basic training manual for enhancing the capacity of various stakeholders (Government authorities, policy makers, practitioners (including farmers) and communities) towards climate and disaster resilient agricultural management.

According to FAO Resilience Framework (FAO, 2016), there are three main groups of shocks that govern agriculture sector which includes natural hazards, food chain crises and protracted crises. Natural hazards include climate related disaster. Food chain crises include avian influenza, locust infestation, food borne pathogens and mycotoxins. Protracted crises are driven by a combination of recurring causes- human made factors and natural hazards.

FAO framework for agricultural resilience is based on four pillars- crisis and disaster risk governance, monitoring and early warning, prevention and vulnerability reduction and emerging preparedness and response.

There are efforts for mainstreaming climate change adaptation and disaster risk reduction strategies.

13.9. Skills shift to ward off factors of vulnerability

Moving to a green and climate-resilient economy requires a dramatic shift in the existing skills of the workforce. According to Greening Technical and Vocational Education and Training: A Practical Guide for Institutions (UNESCO-UNEVOC, 2017), this shift requires changes in the way jobs are performed, including both the development of skills to equip people to work in new climate-friendly occupations and the greening of existing jobs.

Adapting to climate change can be done in different ways. Irrespective of the country and regardless of the occupation, technological orientation acts as an enabling factor for effective climate change adaptation.

Adaptation can be addressed simply through a behavioural action, for example by shifting people from a vulnerable area to a safer area. It can also be dealt with through the formation of new skills and the development of people's capacities to face vulnerabilities. But often skills development needs to be technologically focused because adaptation can also be achieved through employing different forms of technology; for example, introducing a new irrigation system or a climate-resilient seed will be a technological transformation which requires a new set of skills.

The transition towards a green economy requires greening jobs/professions. The content analysis of climate policies and strategies reveals that countries lack a common definition of what a green job is. However, within different contexts, various countries have adopted a range of approaches to achieve a green economy, for example: Green Jobs Act of Philippines, 2016; Decent Work of Canada, ILO's efforts in Turkey and Job Creation of USA.

References

Earth.org 11 ways farmers are adapting to climate change, https://earth.org>11-ways-farmers-are-adapting

Eaton Business School The importance of leadership in a VUCA world, https://ebsedu.org>

FAO 2016 Increasing the resilience of Agriculture livelihoods, https://www.fao.org/3/i5615/i5615cipdf

Gupta A K, Srivastava, R, Hodan, S, Singh, S and Chary, G R 2021 Climate and Disaster Resilient Agriculture Management: Training Manual, National Institute of Disaster Management, New Delhi, India

Hager Group From VUCA to BANI: Instability as a new paradigm, https://hagergroup.com>Home>blog>trends-innovation

Indeed, Resilience skills: Definition and Examples, https://www.indeed.com>career development

Inspire Group Mastering BANI: Learning Design in an uncertain world, https://www.inspiregroup.net>blog>bani-for-learning

PwC Workforce for the Future: The Competing forces shaping 2030, https://www.pwc.com>workforce>publications>workforce-for-the-future

Times Pro Developing Skills to navigate disruption and thrive in a vuca world, https://timespro.com>blog

UNESCO 2021 Skills Development and Climate Change Action Plans: Enhancing TVET's Contribution, UNESCO-UNEVOC, https://unevoc.unesco.org>pub>skills_development

World Skill Council, Research and Innovation https://www.worldskill.org>research_and_innovation

CHAPTER 14

EPILOGUE

This chapter presents a critique of the skilling domain in India, with particular reference to the agricultural sector and suggests how to move ahead to improve the present setbacks.

14.1. Quantity vs quality

A critical analysis of the skill development scenario of India reveals that despite ambitious targets and substantial public investments, the skill development landscape has remained mired in quantity driven metrices and dubious success stories. The fact is that the skill development ecosystem in India is complex, large and diverse, providing varied levels of skills across an extremely heterogeneous population. This poses many challenges and concerns about the skill development programmes.

While formal vocational training has witnessed a modest rise to 4.1% in 2023 (from 1.8% in 2017), informal training avenues-particularly hereditary (from 1.45 in 2017 to 11.6% in 2023) and on-the-job training (from 2.04% in 2017 to 9.3% in 2023) have surged (Sharma and Mehrotra, 2025). The authors highlighted a large increase of workers already in the workforce reported to be imparted skill training after attending one day to one week recognition of prior learning (RPL) exercise, which is a way of certifying those with self-learnt/on the job skills. A high influx of RPL trained individuals is given as a possible explanation for the increased quantity of trained workers within a short span. However, it is observed that there is rapid decrease in the duration of skill development courses, which had definitely affected the quality of trainings.

As we know, skill cannot be acquired within a short time, it needs more time and effort. We are also aware about the failure stories of many technological interventions in agriculture and allied sectors due to lack of proper skill trainings offered to farmers. Skill development is altogether a different process, which needs a thorough hands-on experience and handholding support for the trainees to imbibe the skill. Also, it must be practiced and improved upon for better perfection.

Generally, farmers are given trainings in the name of awareness training, exposure training, technical training, etc about the technologies for a brief period or one day or two and accordingly, the organisers claim that the farmers are equipped to adopt the technologies. This is a sad state of affairs evident in the farm sector.

The National Apprenticeship Promotion Scheme (NAPS) launched in 2016 with a budget of Rs. 10,000 crores was aimed to train 50 lakh apprentices by 2020. However, by 2022, only 20 lakh apprentices had been trained, and only approximately Rs. 650 crore of the allocated funds were disbursed to states between 2017 and 2022.

India replaced NAPS with NAPS 2.0 in August 2023. Some of the changes include a direct benefit transfer from the government to the candidate's bank account. The age limit for those who want to apply under NAPS has been increased to 35.

Under NAPS-2, a budget of Rs 1,942 crore has been allocated for four 4 years from FY 2022-23 to FY 2025-26. The target is to enrol 46 lakh apprentices over a period of four years from FY 2022-23 to FY 2025-26. Though NAPS 2.0 was introduced in 2023, details about the performance are not yet available (Mehrotra and Sharma, 2025).

A 2022 study by ILO noted that the amendments to the Apprenticeship Act of 1961 have resulted in some increase in apprenticeship numbers. Despite this, apprentices still represent a miniscule fraction of India's total workforce, just over half a million apprentices out of a workforce of 570 million in 2022-23, only a modest increase from 2.5 lakh ten years ago.

The various available reports indicate that India's vocational education system has become a number game, where certificates are churned out without meaningful skill acquisition. The rising reliance on short-term training, poor employment outcomes, and inflated placement statistics reflect systemic failures, which need to be addressed. If India truly aims to become a global skill hub, it must prioritise quality over quantity, invest in long term training programmes and increase accountability of private sector in vocational training rather than considering them in existence as a mere spectator.

The vocational training and skill development landscape has been evolving in India from the post-independence era. In recent years the fields of science, technology, engineering, and mathematics, collectively known as STEM, have received much attention for their critical role in maintaining our nation's competitive edge in the global economy. But now, skill development has received both national and global attention.

Skill development at the village level is now undertaken by self-help groups and/or farmer clubs, etc. Most of these players work in partnership with each other and in partnership with the government. However, the training curricula and methods do not always conform the NSQF and hence acceptability of such certification is quite varied. With the implementation of the NSQF, it is targeted that all trainings will have to be NSQF-compliant and training centres and partners need to be accredited and affiliated as per guidelines. This is to be strictly adhered to, which will give a major boost to quality skill training.

14.2. Agricultural Extension vs Skill training

Since 2013, India has been introducing new, nationally standardised agricultural vocational training programmes, as part of the broader Skill India initiative of vocational education reform. However, given a dearth of existing specialised vocational education centres capable of providing agricultural training, India has been relying on other institutions to implement training at the ground level, notably institutions of agricultural extension. This has given rise to several imbalances, as agricultural vocational education and agricultural extension proceed from different assumptions. Agricultural Extension has never functioned as a formal skill training provider, resulting in certification (Brown, 2020). It has always been less formal and more production-oriented focusing on knowledge, rather than skill.

The SSC overseeing the introduction of new TVET projects in agriculture is the Agriculture Skill Council of India (ASCI), which has three key mandates. The first is to identify the skills needs of the agricultural sector and formulate 'qualifications packs'. Second, ASCI is to provide accreditation of agencies capable of delivering training for these 'qualifications packs', ensuring that they have the relevant infrastructure and expertise. Third, ASCI oversees the processes of assessing and certifying skills of the trainees.

Given the lack of existing agricultural TVET institutions, India's agricultural extension institutions became logical partners for the implementation of ASCI's qualifications packs. Extension staff viewed themselves as qualified to implement the ASCI scheme, given their expertise in both agricultural science and strategies of engaging with rural communities. It has become a standard assumption in TVET that trainers should have both industry experience (in this case, experience in field agriculture) and some qualifications in vocational pedagogies. How far these hold good in the case of the present-day trainers of the SAU system is a big question, yet to be researched.

14.3. India's role in Skill Development efforts

India is strategically positioning itself for a global talent revolution, with key initiatives like the National Education Policy (NEP), Skill India Mission, and Digital India driving its commitment to vocational training and skill development. It is expected that by 2025, over 50 per cent of India's secondary and tertiary students will have access to vocational training, ensuring they are aligned with global industry standards. As sectors like AI, cybersecurity, data science, and e-commerce transform industries worldwide, India's emphasis on digital readiness ensures its workforce can seamlessly integrate into the interconnected job market.

India's strategic importance in the global labour market is clear, as its skilled workforce continues to migrate to fill critical gaps across sectors such as IT, healthcare, engineering, and cybersecurity. By embracing global skills mobility, investing in public infrastructure, and aligning its educational systems with the needs of the international job market, India can not only boost its economy through foreign remittances but also solidify its position as a leader in the future of work.

Governments across the world are grappling with the enormous task of providing jobs to their working- age populations. However, creation of jobs is only one side of the coin; the other is the need to address the problem that most people do not have the required skills to perform these jobs. Given the importance of having people with the right skills and matching them to the right employment opportunities, skills development arises as a viable area of intervention. Skills development is a priority for governments and the private sector alike. The private sector stands to gain by investing in skills development.

The rapid advancement of technologies such as artificial intelligence, automation, and digitalization is reshaping industries worldwide. Projections indicate that nearly 25 per cent of all jobs will undergo significant changes within the next five years, necessitating a workforce equipped with adaptable, future-ready skills.

The skill development issue in India is pertinent both at the demand and supply level. Generating employment is definitely a challenge given the enormity of population entering workforce each year. From the supply side, the issue is primarily related to employability of the workforce due to varying reasons ranging from poor education, lack of training facilities, inadequate skilling, quality issues leading to mismatch of skill requirements, and poor perception of vocational skilling vis-à-vis formal education. These have inadvertently created skill shortages and also contributed to higher unemployment. Hence, both employment and employability are key factors of concern today.

With the world's largest youth population, India has an opportunity to ride a demographic tide to social and economic progress. Around 66 percent of the current population is under the age of 35 and the window, in which India can harness its 'demographic dividend' from this youth bulge is expected to remain open until 2040. (Ernest Young & FICCI, 2013).

Skill development initiatives are no longer optional; they are imperative. These initiatives hold the key to unlocking new opportunities, enhancing global competitiveness, fostering innovation, and driving sustainable growth. Digital platforms and edtech solutions are democratizing access to education and skilling, enabling Indian professionals to engage in the global economy without relocating.

14.4. Technology vs Skill development

While technology has been a catalyst for job creation, it also poses significant challenges in terms of job displacement due to automation. On one hand, Indian unicorns and startups have created over one lakh jobs every year since 2019, and new roles have been created like data scientists, AI specialists, and digital marketers. On the other hand, it has been predicted that about 69 per cent of jobs in India are threatened by automation, which is also a global threat. It is estimated that 56 per cent of the salaried employment in Cambodia, Indonesia, Philippines, Thailand, and Vietnam are at high risk of displacement. In India too, it is estimated that by 2030, 9 per cent of our work hours in India could be automated. So, there is a challenge for policymakers to leverage the job-creating potential of technology while mitigating its adverse impact.

The National Career Service (NCS) was set up in 2015 by the Ministry of Labour and Employment to provide a platform for job seekers and employers to fulfil their requirement for gainful employment. The platform provides jobs postings, upcoming job fair notifications, and also skilling programmes for jobseekers.

The current state of the world is characterised by technological disruptions, global realignments and environmental pressures. The organizations are moving toward a whole new operating model for work and the workforce that places skills, more than jobs, at the centre. We call this new operating model for work and the workforce 'the skills-based organization.'

Refocusing work around the people doing it and the skills required to do so, and supplying the necessary skills training can enhance the employability. For example, identifying adjacent or foundational skills of workers displaced by automation or whose roles are no longer needed can help organizations redeploy them to work that is needed.

Upskilling and reskilling should be among governments' key priorities in the context of evolving skill needs and changing labour markets. As highlighted by the first principle of the European Pillar of Social Rights, "everyone has the right to quality and inclusive education, training and lifelong learning in order to maintain and acquire skills that allow full participation in society and successful transitions in the labour market" (European Commission, 2017). (Annexure-10)

14.5. Challenges to skills development

While the targets are ambitious, the skills development space in India is also facing certain barriers that require the combined expertise, know-how and resources of all stakeholders to address these issues at scale. India's eventual success in skilling the millions will depend on formulating and disseminating innovative solutions to these challenges.

The most important handicap of skilling programmes in general has been our inability to precisely assess their impact in employment generation, and the gap in data as well as institutional mechanisms to address the work trajectories of trainees' post training. The collection of longitudinal data pertaining to the impact of skilling programmes across state, private and NGO platforms present challenges to our efforts to bolster the skilling environment.

The main problems identified in implementation of skilling programmes in comparative studies are primarily at two levels - questions of provision of skill training and questions of integration into the job market after training (Mythreyi *et al* 2019).

The challenges are four-pronged. First, the world faces major job market transformations. Technological progress and AI are creating a structural labour market churn, reshaping the job landscape, and raising concerns about job displacement. Second, we face significant demographic concerns. The ageing global population presents a significant challenge to maintaining economic growth. Third, we will face increased environmentally driven migration. Migration triggered by environmental pressures could shift future carbon emissions patterns. Finally, numerous policy dilemmas will be caused by advances in Gen-AI. The rapid pace of AI progression is creating numerous policy challenges including regulatory issues, AI's carbon footprint, ethical considerations, security risks, and potential job losses.

Sustainability should be viewed broadly as encompassing human well being rather than just an economic activity. Generative AI requires a comprehensive regulatory framework, ethical guidelines, and proactive industry self-governance.

The trends in skill development should focus on:

1. Embracing human sustainability: Many organisations may be stuck in a legacy mindset that centres on extracting value from people rather than working to create value for them. The perspective of the organisation needs to be oriented around the idea of human sustainability.
2. Moving beyond productivity to measure human performance: Traditional methods of measuring work productivity as a series of inputs and outputs reflect the perspective of the organisation. New approaches should consider the worker as a human being, with a more nuanced perspective as to how he/she contributes to the organisation.
3. Balancing privacy with transparency to build trust: There is significant potential for misuse of transparency such as privacy breaches, AI driven surveillance and efforts to control the movements of the workers. If responsibly managed, transparency can create new opportunities to measure and unlock human performance.
4. Overcoming the imagination deficit: Technological disruption is outpacing the capacity of many organisations and workers to imagine new ways of working that gets the best out of both humans and technology. Consequently, many organisations may soon be facing an imagination deficit.
5. Creating digital playgrounds to explore, experiment and play: There is a growing demand for safe spaces in which both organisations and individuals can imagine, explore and co-create a future that delivers better

human experiences and outcomes at speed and scale. Deloitte calls these spaces 'digital playgrounds'.

6. Cultivating workplace microcultures: According to conventional wisdom, corporate culture should be one size-fits-all, that ensures everyone is working in the same way. However, in reality, organisations typically consist of a diverse set of microcultures. A key to harnessing the power of microcultures is encouraging the autonomy of various work groups.
7. Making the shift to boundaryless HR: Work is increasingly demanding agility, innovation and collaboration to achieve outcomes. Boundaryless HR can develop people-discipline expertise and weave it throughout the fabric of business, creating multidisciplinary solutions. 8.The Indian gig economy, including its agriculture segment, is booming with platforms facilitating work in areas like farming, agro processing and delivery services and is expected to create enormous job opportunities and contribute significantly to India's GDP.

14.6. Future roadmap

- Innovation, livelihood and entrepreneurship should be made integral to skill development in agriculture. All skill development programmes, whether part of the vocational education and training or formal education, should emphasize innovation and entrepreneurship and linked to livelihood opportunities.
- Since the focus is shifting from Employability to Entrepreneurship, there is a need to shift the implementation approach of agencies to work towards developing entrepreneurship skills. Youth, especially those engaged in the agriculture sector must be targeted.
- Expand the frontiers of skill development in the agriculture and food sector. Targeted beneficiaries with respect to skill development in the agricultural sector comprise two distinct categories. First, youth trained on different job roles in agriculture can venture into the sectors as practitioners and/ or extension and technical advisory workers. Second, existing farmers can benefit from upgrading their skill sets towards efficient farming practices.
- Recognition of prior learning (RPL) should be properly assessed and upskilling need to be done for creating confidence in the farmers and for more productive work environment.
- Curriculum design, examination, assessment and certification processes should be structured in consultation with the industry and sectoral experts to bring in the market requirements. Successful agricultural entrepreneurs should form part of this consultative process and training programmes as experts and mentors.
- Fostering skill development through firm-farm linkages can help encourage private sector participation in skill development in the agriculture and food sector.

- To overcome the imagination deficit due to technological disruptions, organisations need to scale and operationalize the increased use of distinctly human capabilities, such as curiosity, empathy, creativity, etc and the workers and teams should be given the autonomy to use these capabilities to shape the kinds of work they do.
- In the absence of a dedicated faculty for providing skill trainings in the agriculture sector, there is need for organising appropriate 'training of trainers' programmes jointly by ICAR and ASCI in partnership with MANAGE, NAARM and ATARI for enhancing the quality of skill providers.
- Opportunities should be provided for gig workers in agriculture for upskilling and reskilling and ensuring social security safeguards.
- Strive towards sustainability of skill development programmes. The Government of India has sanctioned massive outlays to jumpstart the revamped skill development programme and make it more attractive and affordable to the masses. However, in moving forward, it is important to ensure that the programme is financially viable. To make this happen, the programme should be demand driven and linked to job opportunities across sectors, whereby people having undergone the skill development programmes are able to find suitable jobs. This will motivate private players to join the skilling programme realising the benefit.

References

Brown, T 2020 Pathways to Agricultural Skill Development in the Indian Himalayas, Journal of South Asia Development 15 (2): 270-292

Deloitte What do organisations need most in a disrupted, boundaryless age? More imagination, https://www2.deloitte.com>focus>human-capital-trends

Ernst Young and FICCI 2013 Knowledge Paper on Skill Development in India-Learner first, https://imyriaads.files.wordpress.com>2013/02

ILO 2022 Good Practices in Apprenticeships in India: Challenges and Opportunities, https://researchrepository.ilo.org

Maithreyi, R., Shruti Padmanabhan, Niveditha Menon & Jyotsna Jha. 2019. A Critical Sociological Analysis of the Skills Development Initiative of India. Bengaluru: Centre for Budget and Policy Studies, available at http://cbps.in/wp-content/ uploads/Final-Skills-Full-Report_19-July-2019.pdf

Mehrotra, S and Sharma, H 2025 The Illusion of Skill development in India: Decoding the sudden increase in the 'Vocationally Trained', https://m.thewire.in>article>labour>the-illusion-of-skill

Sharma, H and Mehrotra, S 2025 Skill Development in India: The facts behind the figures, https://idronline.org>livelihoods

Wheebox 2025 India Skill Report 2025: Unlocking Indian skills for the world, https://wheebox.com/assets/pdf/ISR_Report 2025

ANNEXURE 1

Need for a System of National Skill Universities (NSUs)

The NSQF offers a holistic qualification framework for skills education across educational sectors, educational and training institutions, and industry. However, skills administration under NSQF has not rolled out as expected. The factors that have impeded the roll-out include:

1. Absence of University based systems for recognition of prior learning and a competency based modular course and credit framework across educational streams.
2. Absence of courses designed to meet the needs of specific industry sectors. Industry linked Universities dedicated to industry-oriented learning are needed.
3. Absence of dedicated educational institutions to develop an array of NSQF-compliant curriculum packages across sectors.
4. Lack of integration of skills training in industry with the University systems. Germany and Australia , the world leaders in skill education have effectively gone beyond industry linkage and integrated skill education with their higher educational system.
5. Lack of alignment of existing Universities with industry needs.
6. Absence of Centres of Excellence in skills education. Vocational education in India has traditionally been viewed as appropriate for workers at the bottom of the work pyramid. This perception needs to be changed.
7. Creation of nation-wide system for transfer of credits.
8. Segmentation of skills education across education and training streams and programmes.
9. Investing in skills education is perceived as involving high risk in investment made by learners, educational institutions and banks alike.
10. The academic institutional models necessary to operating NSQF do not exist and need to be seeded and evolved.

Source: The National Skill Universities Bill, 2015,

ANNEXURE 2

Skill Universities in India

Sl. No	Name of the Skill University	Year of establishment
1.	Team Lease Skills University, Vadodara, Gujarat (Pvt))	2013
2.	Seacom Skills University, Santiniketan, West Bengal (Pvt)	2014
3.	Shri. Vishwakarma Skill University, Palwal, Haryana (State)	2016
4.	Symbiosis Skills and Professional University, Pune, MS (Deemed)	2017
5.	Bharatiya Skill Development University, Jaipur, Rajasthan (Pvt)	2017
6.	Rajasthan ILD Skills University (RISU), Jaipur, Rajasthan (State)	2017
7.	Delhi Skill and Entrepreneurship University, Dwaraka, New Delhi (State)	2020
8.	Lamnn Tech Skills University, Chandigarh, Punjab (Pvt)	2021
9.	Maharashtra State Skills University, Mumbai, MS (State)	2021
10.	Atlas Skilltech University, Mumbai, MS (Pvt)	2021
11.	Kaushalya Skills University, Mahatma Gandhi Labour Institute (MGLI), Ahmedabad, Gujarat	2021
12.	Medhavi Skills University, Sikkim (Pvt)	2022
13.	Assam Skill University (funded by ADB)	2024

Source: https://www.nirfindia.org>Rankings>SkillUniversities

ANNEXURE 3

Sector Skill Councils (SSC)

Sector Skill Councils are set up as autonomous industry-led bodies by NSDC. They create Occupational Standards and Qualification bodies, develop competency framework, conduct Train the Trainer Programs, conduct skill gap studies and Assess and Certify trainees on the curriculum aligned to National Occupational Standards developed by them. NSDC is mandated to initiate and incubate SSCs with initial seed funding to facilitate their growth and enable them to achieve self-sustainability in a time bound manner.

The SSCs have been mandated with the following functions:

- **Identification of skill development needs** including preparing a catalogue of types of skills, range and depth of skills to facilitate individuals to choose from them.
- Development of a sector skill development plan and maintaining skill inventory.
- **Determining skills/competency standards and qualifications** and getting them notified as per NSQF.
- **Standardization of affiliation, accreditation, examination and certification** process in accordance with NSQF as determined by NSQC.
- May also conduct skill-based **assessment and certification for QP /NOS aligned training** programmes.
- Participation in the setting up of affiliation, accreditation, examination and certification norms for their respective sectors.
- Plan and facilitate the execution of **Training of Trainers** along with NSDC and states.
- Promotion of academies of excellence.
- Paying particular attention to the skilling needs of ST/SC, differently-abled and minority groups.
- Ensuring that the persons trained and skilled in accordance with the norms laid down are assured of employment at decent wages.

There are 37 SSCs as below:

1. Aerospace and Aviation Sector Skill Council

2. Agriculture Skill Council of India (ASCI)
3. Apparel made ups & Home furnishing Sector Skill Council
4. Automotive Skill Development Council (ASDC)
5. Beauty and Wellness Sector Skill Council (B&WSSC)
6. BFSI Sector Skill Council of India (BFSI)
7. Capital Goods Skill Council (CGSC)
8. Construction Skill Development Council of India
9. Domestic Workers Sector Skill Council (DWSSC)
10. Electronics Sector Skill Council of India
11. Food Industry Capacity and Skill Initiative (FICSI)
12. Furniture and Fittings Skill Council
13. Gem and Jewellery Skill Council of India
14. Handicrafts and Carpet Sector Skill Council
15. Healthcare Sector Skill Council
16. Hydrocarbon Sector Skill Council
17. Indian Iron and Steel Sector Skill Council
18. Infrastructure Equipment Skill Council (IESC)
19. Instrumentation Automation Surveillance and Communication Sector Skill Council
20. IT-ITeS Sector Skill Council (NASSCOM)
21. Leather Sector Skill Council
22. Life Science Sector Skill Development Council
23. Logistics Sector Skill Council
24. Management and Entrepreneurship and Professional Skills Council (MEPSC)
25. Media and Entertainment Skill Council
26. Paints and Coatings Skill Council
27. Power Sector Skill Council
28. Retailer Association's Skill Council of India (RASCI)
29. Rubber, Chemical and Petrochemical Skill Development Council
30. Skill Council for Green Jobs (SCGJ)
31. Skill Council for Mining Sector
32. Skill Council for Persons with Disability
33. Sports, Physical Education, Fitness and Leisure Skills Council

34. Telecom Sector Skill Council
35. Textile Sector Skill Council
36. Tourism and Hospitality Skill Council
37. Water Management and Plumbing Skill Council

Source: https://www.msde.gov.in>organizations>ssc

ANNEXURE 4

Agri Clinics and Agri Business Centres (AC-ABC)

The Ministry of Agriculture and Farmers Welfare, Government of India, in association with NABARD has launched a unique programme to take better methods of farming to each and every farmer across the country.

This programme aims to tap the expertise available in the large pool of Agriculture Graduates. Irrespective of whether being a fresh graduate or not, or whether currently employed or not, one can set up own AgriClinic or AgriBusiness Centre and offer professional extension services to farmers.

Committed to this programme, the Government is now providing start-up training to graduates in Agriculture, or any subject allied to Agriculture like Horticulture, Sericulture, Veterinary Sciences, Forestry, Dairy, Poultry Farming, and Fisheries, etc. Those completing the training can apply for special start-up loans for venture.

Earn money and prestige by becoming a consultant to farmers

Agribusiness Centres would provide paid services for enhancement of agriculture production and income of farmers. The Centres would need to advice farmers on crop selection, best farm practices, post-harvest value-added options, key agricultural information (including perhaps even Internet-based weather forecast), price trends, market news, risk mitigation and crop insurance, credit and input access, as well as critical sanitary and phyto-sanitary considerations, which the farmers have to keep in mind.

Free Training to set up your Agriclinic or Agribusiness Centre

As an integral part of this nationwide initiative, specialised training will be provided to Agriculture Graduates interested in setting up such a centre. Being provided free of cost, 45 days training course will be offered by select institutes across the country. Initiated by SFAC, and co-ordinated by MANAGE, the course comprises Entrepreneurship and Business Management, as well as skill improvement modules in one's chosen areas of activity.

Bank loans available for Agriclinics and Agribusiness Centres

Ceiling of project cost for subsidy has been enhanced to Rs.20 lakhs for an individual project (25 lakhs in case of extremely successful individual projects) and to Rs.100 lakhs for a group project. Depending on the type of venture one want to set up, and with a moratorium of up to 2 years, Agriclinics and Agribusiness Centre Loans can be repaid within 5 to 10 years as per easy installment plans. The rate of interest, margin and security on loans will be decided by the respective bank, as per RBI norms. Depending on one's entitlement, one can even apply for margin money assistance.

Source: https://www.nabard.org>content1

ANNEXURE 5

Three Operational Models for Centre of Excellence

Centre of Excellence (CoE) is a body that provides leadership, best practices, research, support, training of trainers and skill training for a specific sector/s. CoE in skilling ecosystem is envisioned to be one stop resource centre, established/working in partnership with industry to raise training standards, boost productivity, address emerging skill gaps and align training & research with industry needs.

With the intent to overcome skill demand-supply mismatch, to have continuous supply of skilled workforce and disseminate best practices, "Centre of Excellence" is recognized by Ministry of Skill Development & Entrepreneurship (MSDE). The initiative would encourage such bodies already engaged in research and development activities in the skilling domain and allied fields to work on key emerging areas where there is knowledge deficit or skill gap, so as to set up Centres of Excellence.

Largely, three models are existing in the skilling ecosystem for operating-"Centre of Excellence", as below:

Model 1: Centres within a training institution

The most common is establishing a centre of excellence within an institution, such as a university (especially existing skill universities) or training provider. Existing facilities may be used, or new ones may be constructed. The institution once declared as a CoE can be elevated from "good" to "great" by acquiring expert/ master trainers, upgrading existing facilities, acquiring tools, equipment and facilities as per the market requirement, introducing domain-specific blended learning. Typically, this centre will become a model of best practice in the designated field. This approach enables the new centre to take advantage of the reputation and resources of an existing institution with an established reputation in the skill domain. The CoE could partner with various industries, government, sector skill councils, and industry bodies to upscale its competency and the existing Skill Universities may adopt this model if they meet the criteria for recognition.

One of the examples in this case is Centre of Public Policy (CoE) at IIMB. It is an independent public interest-oriented policy think tank engaged in pioneering research, teaching, training and capacity building.

Model 2: Stand-alone independent centres

In some situations, a centre of excellence is set up from scratch on a brownfield or greenfield site not connected to an existing institution. This requires intensive capital and human resources investment, as all facilities, recruitment, operational structures, and processes need to be established. This may take several years, so a positive reputation needs to be quickly established. Involving industry in co-financing and sponsoring the centre can be highly effective. These centres may be linked to wider governance or organizational structures.

For example, a stand-alone centre may be established by an industry skills council that has identified training needs that cannot be addressed by existing providers. Example in this case is the Tata Institute of Social Sciences and UNFPA have launched a Centre of Excellence on Youth and Adolescents.

Model 3: Networks of excellence

These are organizational structures or agencies that bring existing training providers together in a network. This can be an appropriate option when good-quality training providers operate in isolation from each other or do not collaborate with industry.

Establishing a network does not involve the physical construction of new training facilities. Capacity-building efforts are focused on existing providers. These may involve upgrading equipment, strengthening human resources, and the granting of a recognized "kitemark" or certification of quality.

Networks of excellence can be an effective means of ensuring that provision in specific areas is aligned with government policy priorities. They are also particularly useful in situations wherein the skills needs of a particular sector or subsector are diverse.

Example: CII Institute of Logistics, Chennai, provides specialized Services to build Leadership, Strategy for Sustainability, Identify and Adapt Clean Technologies and Management, Enhance Energy efficiency and become Environment friendly, adopt world class practices in Logistics and Manufacturing Management.

Source: Guidelines for Recognition of 'Centre of Excellence', 2019., Ministry of Skill Development and Entrepreneurship, Government of India

ANNEXURE 6

Ten new leadership skills for an uncertain world

This book introduces ten new leadership skills for the future. They are:

1. Maker instinct: Ability to exploit inner drive to build and grow, as well as connect with others in the making. Leaders need this basic skill to make and remake organisations.
2. Clarity: Ability to see through messes and contradictions to a future that others cannot yet see and appreciate. Leaders must be clear about what they are making, but flexible about how it gets made.
3. Dilemma flipping: Ability to turn dilemmas into advantages and opportunities
4. Immersive learning ability: Ability to immerse oneself in unfamiliar environments, to learn from them in a first person way.
5. Bio-empathy: Ability to see things from nature's point of view, to understand, respect and learn from its patterns
6. Constructive depolarizing: Ability to calm tense situations where differences dominate and communication has broken down and bring people from divergent cultures towards positive engagement.
7. Quiet transparency: Ability to be open and authentic about what matters, without being overly self-promoting
8. Rapid prototyping: Ability to create quick early versions of innovations with the expectation that later success will require early failures
9. Smart -mob organising: Ability to create, engage with and nurture purposeful business or social change networks through intelligent use of electronic media and in-person communication
10. Commons creating: Ability to seed, nurture and grow shared assets that can benefit all players and allow competition at a higher level. This is the most important future leadership skill.

*Source: Bob Johansen, **Leaders Make the Future:** Ten New Leadership skills for an uncertain world, Berrett-Koehler Publishers, Oakland, USA*

ANNEXURE 7

Report of the Working Group on Knowledge Economy, Employment & Skills Kerala State Planning Board (14th FYP)

Need of new skills in technology:

Worldwide data indicate that there is a decline in demand for medium-skilled workers. This is particularly important because most young people entering the labour force are actually at the middle skill level, so it requires a rethinking of skill needs. There is a growth in the adoption of robots in manufacturing. This has implications in the long run for new kinds of skills which are more in tune with the new technology.

New opportunities and lifelong learning:

In accordance with the changing world of work, outlined by the ILO Global Commission, people need to be equipped with skills for the job opportunities emerging from new sectors, including the (a) digital economy; (b) green economy; (c) care economy; and (d) rural economy. The rural economy will continue to play an important role in job creation. The new skill eco system sees lifelong learning (LLL) as a guiding principle, supporting people.

Skilling in clusters:

Formal industry in Kerala could be engaged in skill development in agro-processing, bamboo, coir, cashew, and MSME clusters that are doing well. Support could also be sought from the National Skill Development Corporation to start skilling centres near the clusters.

High Touch High Tech (HTHT) Education With the Assistance of Artificial Intelligence (AI):

A new learning system combining high technology with classroom teaching (human touch) changed the way of teaching and learning in South Korea, especially after Covid-19, from one of mass standardisation to one of mass personalization and lessons from such experiments should be incorporated into the educational system in the state.

Employment in Unorganised Sector

Unorganised sector plays a crucial role in the Indian economy in terms of employment and its contribution to the Gross Domestic Product, savings and capital formation. More than 90 per cent of workforce and nearly 50 per cent of the

Gross Domestic Product is attributable to this sector. A high proportion of socially and economically weaker sections of society are engaged in the unorganised economic activities in India and Kerala.

The schemes on skill development and employment are spread across various sectors and implemented by various departments. Most of these schemes are multi-purpose in nature that covers wage employment, self-employment, training and skill development activities also.

The COVID-19 pandemic has seriously disrupted the skilling, up skilling and reskilling of employees, apprentices and interns in all types of enterprises and organizations globally and particularly in Kerala during 2020. Millions of people were lost jobs, and others rapidly adjusted to working from home as offices closed. Many other workers were deemed essential and continued to work in hospitals and grocery stores, on garbage trucks and in warehouses, yet under new protocols to reduce the spread of the novel coronavirus.

In order to enhance the agility and resilience of societies to deal with the challenges posed by the pandemic and crises in the future, governments have a crucial responsibility in ensuring the continuation of training and development at workplaces, as well as improving the effectiveness of skills development and lifelong learning systems.

While the higher education segment across the country has a strong focus on knowledge, the focus on skills has been relatively low. The gap was predominantly closed by Skills and Training Service Providers.

The Digital University and K-DISC have initiated various interventions in this on the technology infrastructure side.

- Kerala Food Platform for Safe to Eat Food with traceability implemented in Palliyakkal Cooperative Bank with Proof of Concepts (PoCs) for traceability, Participatory Guarantee System, e-commerce and Green Army systems.
- Carbon Neutral Wayanad for promoting sustainable cultivation of coffee
- Specific proposals from Kerala Agriculture University (KAU), Kerala Veterinary University (KVASU) and Kerala Fisheries University (KUFOS) to undertake Proof of Concepts (PoCs) and pilots pertaining, but not limited, to cutting edge areas like: Precision Agriculture, Drone-based Agriculture, Smart Livestock Farming, Precision Feeding, Animal Health Monitoring, Precision Aquaculture

Gig Economy

Gig-based employment can create significant incremental economic value and increase livelihood opportunities for low-income workers. The gig economy benefits employers, employees, as well as the economy at large with gains extending beyond the traditional notions of convenience and on-demand availability and flexibility. Gig work is commonly understood as temporary on-demand labour employed in a task-based or a time-based manner.

The following are the four distinct, critical, and inter-related components to the gig ecosystem.

1. Platforms: Providing gig services to end-users and addressing industry and worker specific prerequisites, and innovations in offerings, pricing, and delivery models.
2. Ecosystem Enablers: Providing supporting and commonly used services across platforms, including background verifications, financial services, training and skill development.
3. Digital Public Goods: Providing a set of common, secured, and consent-based digital goods serving as 'community-owned' systems-of-records and interoperability frameworks to enable portability and transparency of demand and supply-side information.
4. Public Policy and Regulations: Governing the rules of the road for the gig economy, covering worker protection, social benefits, and standard of care.

Source: https://spb.kerala.gov.in>sites>files>inline-files

ANNEXURE 8

Report on National Programme on Artificial Intelligence (NPAI) Skilling Framework

1. The National Programme on AI Skilling Framework report was approved in 2023 by MSDE and is aimed at equipping the nation with the necessary skills in Artificial Intelligence (AI) including Generative AI.
2. As per recommendation, the skilling requirements in AI shall be categorized under three broad categories:
   a. For "All": for the end users who require basic digital literacy skills to effectively use AI interfaces, & tools, understand responsible AI practices, and navigate potential pitfalls
   b. For "Many": This encompasses AI-empowered individuals who seek digital fluency to enhance productivity, utilize AI tools, and grasp AI applications in their respective fields.
   c. For "Few": AI-skilled professionals who possess digital mastery, encompassing technical expertise in areas like coding, data analytics, machine learning, cybersecurity, and cloud computing, information security and other related technologies, enabling them to develop AI products and solutions across domains. This also includes specialized new areas such as Explainable AI, AI in legal & legal for AI, which will be created.
3. The report provides high level details of the courses that are existing and the ones that may be created to meet the future demand, both in IT industry and the other business verticals This will involve establishing AI-focused training institutions, creating partnerships with industry leaders, fostering research and development, and promoting entrepreneurship in the AI domain, together with need to Collaborate with Ed-Tech partners.

Some of the areas covered are:

   a. Introduction to AI should start from the early school levels where children can be taught AI tools and be introduced to what AI is and recognize where it is being used.
   b. There is a large amount of course material available from various sources Along with creation, reuse of existing content should be encouraged and multiple modes like self- paced and blended, based on the audience can be leveraged

c. Every course, small or big, must have a module on Ethical AI for a minimum of 10% of its duration Ethical considerations, transparency, fairness, and privacy must be integrated into AI training programs to ensure that AI systems are developed and deployed responsibly.

Source: https://www.ugc.gov.in>pdfnews

ANNEXURE 9

Indo-German Global Academy for Agroecology Research & Learning (IGGAARL)

IGGAARL was established in 2022, in Pulivendula, Kadapa, which is the first of its kind in India. It is an institute which focuses on democratizing science, by ensuring that farmers are key stakeholders in the scientific process.

RySS, in collaboration with the Government of Germany and Government of Andhra Pradesh, is the Project Executive Agent for IGGAARL. The Indo-German Academy conducts research on agroecology and will disseminate knowledge by training and capacity building for the farmers.

The Academy has four thematic pillars: i) Research and Development, ii) Learning and Education, iii) Training and Capacity Building and iv) Up-Scaling. The work across the thematic pillars (given below) will take place across different hubs, in the state of Andhra Pradesh.

1. Farmer-centric and farmer-driven research
2. Agro-ecology standards, testing protocols and centres
3. Agroecology learning, knowledge research and consultancy services
4. Both on-farm and off-farm labs for primary research
5. Research in areas of carbon sequestration, socio-economic domains of agroecology and anthropology, development, agroecology and pest management, agroecology and biogeochemical cycles, value chains and transformations, market linkages, economics of agroecology, sustainability of agroecology etc.
6. Create a digital repository of data and information from the academy
7. Ensuring peer reviewed publications in Journals of prominence by scientists and agroecology practitioners
8. Ensuring Global Knowledge Platforms through the Centre

IGGAARL will establish 'climate resilient villages' through capacity building of thousands of farmer scientists and appointment of one mentor for every 5 farmer scientists in one RBK.

Source: https://apcnf.in>academy

ANNEXURE 10

European Pillar of Social Rights

The 20 Principles of the European Pillar of Social Rights are the beacon guiding towards a strong Social Europe that is fair, inclusive and full of opportunity.

1. Education, Training and life-long learning
2. Gender equality
3. Equal opportunities
4. Active support to employment
5. Secure and adaptable employment
6. Wages
7. Information about employment conditions and protection in case of dismissals
8. Social dialogue and involvement of workers
9. Work-life-balance
10. Healthy, safe and well adapted work environment and data protection
11. Childcare and support to children
12. Social protection
13. Unemployment benefits
14. Minimum income
15. Old age income and pension
16. Healthcare
17. Inclusion of people with disabilities
18. Long term care
19. Housing and assistance for the homeless
20. Access to essential services

Source: https://employment-social-affairs.ec.europa.au>european-pillar

INDEX